Enzymes in Health and Disease

Inaugural Scientific Meeting of the International Society for Clinical Enzymology
London, September 19–21, 1977

Enzymes in Health and Disease

Editors
David M. Goldberg, Toronto, Ont., and J. Henry Wilkinson†, London

61 figures and 55 tables, 1978

S. Karger · Basel · München · Paris · London · New York · Sydney

National Library of Medicine Cataloging in Publication
International Society for Clinical Enzymology, London, 1977
Enzymes in health and disease: inaugural scientific meeting of the International Society for Clinical Enzymology...
Editors: David M. Goldberg and J. Henry Wilkinson
Basel; New York: Karger, c1978
Sponsored by International Federation of Clinical Chemistry
1. Enzymes – congresses 2. Enzyme Tests – congresses I. Goldberg, David M., ed. II. Wilkinson, John Henry, ed. III. International Federation of Clinical Chemistry IV. Title
QY 490 I61e 1977
ISBN 3–8055–2853–1

Printed in Switzerland by Buchdruckerei 'Der Bund', Bern
ISBN 3–8055–2853–1

Contents

Enzyme Standardization and Laboratory Management

Enzyme Methods

Enzymes in Liver Disease

Enzymes in Myocardial Infarction

Enzymes in Kidney Disease

Enzymes in Other Clinical Situations

Genetic Aspects of Enzymes

Abbreviations

The following abbreviations sometimes appear in the text without definition. All others are defined by the authors concerned, or are standard abbreviations accepted by the International Union of Biochemistry.

Enzymes

ALT	alanine aminotransferase (EC 2.6.1.2, L-alanine: 2-oxoglutarate aminotransferase)
AP, ALP	alkaline phosphatase (EC 3.1.3.1, orthophosphoric monoster phosphohydrolase)
AST, ASAT, GOT	aspartate aminotransferase (EC 2.6.1.1, L-aspartate: 2-oxoglutarate aminotransferase)
CK	creatine kinase (EC 2.7.3.2, ATP: creatine N-phosphotransferase)
GGT, γ-GT	γ-glutamyltransferase (EC 2.3.2.2, γ-glutamylpeptide: amino acid γ-glutamyltransferase)
GDH, GLDH	glutamate dehydrogenase (EC 1.4.1.2, L-glutamate: NAD oxidoreductase [deaminating])
G6PD	glucose-6-phosphate dehydrogenase (EC 1.1.1.49, D-glucose-6-phosphate: $NADP^+$ 1-oxidoreductase)
HBDH	α-hydroxybutyrate dehydrogenase (EC unassigned)
Hex	hexosaminidase (EC 3.2.1.30, 2-acetamido-2-deoxy-β-D-glucoside acetamidodeoxyglucohydrolase)
LAP	leucine aminopeptidase (EC 3.4.11.1, α-aminoacyl-peptide hydrolase)
LD, LDH	lactate dehydrogenase (EC 1.1.1.27, L-lactate: NAD^+ oxidoreductase)
MDH	malate dehydrogenase (EC 1.1.1.37, L-malate: NAD^+ oxidoreductase)

Others

BSP	bromosulfthalein
CNS	central nervous system
ECG	electrocardiogram (graphic)
EPE	expert panel on enzymes
GSH	glutathione, reduced
IFCC	International Federation of Clinical Chemistry
NAC	N-acetylcysteine
PBI	protein-bound iodine
PCMB	p-chloro-mercuribenzoate
RBC	red blood cells
SFBC	Société Française de Biologie Clinique
WBC	white blood cells

Introduction

The International Society for Clinical Enzymology was founded at a meeting held in Venice in April 1976 during the Seventh International Symposium on Clinical Enzymology. Its inaugural scientific meeting, the proceedings of which form the subject of the present volume, was held in September 1977 at Charing Cross Hospital Medical School, London, by kind permission of the Council of the Medical School. The conference was attended by about 200 participants from no less than 22 different countries, thus fully justifying use of the adjective 'International'.

The scientific programme was preceded by a short opening ceremony when members were welcomed by Mr. L. C. GRAHAM-DIXON, QC, Chairman of Council of Charing Cross Hospital Medical School, and good wishes for the success of the meeting and the future of the Society were expressed by Dr. J. C. A. RAISON, Chief Scientific Officer of the Department of Health and Social Security, and Dr. J. FREI, President of the International Federation of Clinical Chemistry, under whose sponsorship the meeting was held.

In a three-day conference it is not possible to include all topics of interest and the Organizing Committee decided to restrict the programme to the pathophysiology of enzymes, methodology and standardization, cell enzymes, clinical interpretations, and approaches to enzyme therapy.

The Organizing Committee, which consisted of Professor D. N. BARON, Dr. G. R. DORAN (Secretary-Treasurer), Dr. P. J. N. HOWORTH (representing the Association of Clinical Biochemists), Professor D. W. MOSS, Dr. S. B. ROSALKI and Professor J. H. WILKINSON (Chairman), would like to thank all the contributors to the present volume. They are especially indebted to Mrs. ANN CATER and Mrs. MARJ FLEMING for their valuable secretarial services before, during and after the meeting. Support from the following companies is also gratefully acknowledged: BDH Chemicals Ltd., The Boehringer Corporation (London) Ltd., Calbiochem Ltd., Coulter Electronics Ltd., LKB Instruments Ltd., Fisons Ltd., Pye-Unicam Ltd., G. D. Searle & Co. Ltd., Smith Kline Instrument Co. Ltd., Technicon Instrument Co. Ltd., and Union-Carbide (UK) Ltd.

D. M. Goldberg
J. H. Wilkinson

John Henry Wilkinson (1915–1977)

On November 29th, 1977, JOHN HENRY WILKINSON, Professor of Chemical Pathology at Charing Cross Hospital Medical School and President of the International Society for Clinical Enzymology, died unexpectedly. Known throughout the scientific community for his many original papers, and in particular for his treatises: 'An Introduction to Diagnostic Enzymology' (1962), 'Isoenzymes' (1965 and 1970), and 'Principles and Practice of Clinical Enzymology' (1976), Professor WILKINSON first entered the academic scene in 1947 through an appointment at the Department of Chemical Pathology of the Westminster Hospital Medical School. There, during an 18-year period, he rose to be appointed Reader and added the Doctorate in Science to the Doctorate of Philosophy which he already held.

He crossed the Atlantic in 1965 to take up the Chair of Clinical Chemistry in the University of Pennsylvania, but returned in 1969 to the position he held until his decease. During his busy scientific career, he held the following offices: President of the Association of Clinical Biochemists of the United Kingdom (1973); President of the Pathology Section of the Royal Society of Medicine (1977); and Member of the Editorial Boards of the journals *Clinica Chimica Acta, Enzyme, and Clinical Chemistry.* Among the many honours he received, mention should be made of the following: 'Clinical Chemist of the Year Award' (American Association of Clinical Chemists, 1970), 'Wellcome Prize' (Association of Clinical Biochemists, 1974), and the 'Gianfranco Galanti Award' (1976). He was honoured by the Fellowships of the Royal Institute of Chemistry and the Royal College of Pathologists, but his crowning achievement was his election by Her Majesty to the Commandership of the British Empire in 1977. His loss will be mourned by his family, friends, and colleagues, and not least by the Members of the International Society for Clinical Enzymology to whom he had provided outstanding leadership since its inception.

Enzymes in Health and Disease. Inaug. Scient. Meet. Int. Soc. Clin. Enzymol., London 1977, pp. 1–7 (Karger, Basel 1978)

Plasma Enzyme Activities[1]

J. H. WILKINSON

Department of Chemical Pathology, Charing Cross Hospital Medical School, London

Enzyme tests have undoubtedly proved of immense value as aids to the diagnosis of disease, but little is known of the factors governing the release of enzymes from damaged cells and tissues or the disappearance of enzymes from the circulating plasma. Recent developments in these fields are summarized in this paper.

The Release of Intracellular Enzymes from Tissues

It has long been known that intracellular enzymes may be released into the extracellular fluid and hence into the circulating plasma, not only from necrotic tissues but also in reversible inflammatory disorders. Early studies by ZIERLER [26] showed the release of aldolase from rat muscles to be increased by deprivation of glucose, anoxia, high [K^+] concentrations, and the presence of metabolic inhibitors, from which it appears that the permeability of the cell membrane to intracellular enzymes is controlled by the energy content of the cells. Evidence in support of this theory is provided by the marked increase in the rate of efflux of enzymes from incubated human erythrocytes which occurs after the glucose concentration has fallen to negligible levels [19], and by the reduction of enzyme leakage brought about by the incorporation of ATP into the media in which rat lymphocytes or human leucocytes are incubated [5, 24]. ATP has also been demonstrated to exert a protective effect in human erythrocytes under a variety of conditions [20, 25].

[1] Summary of Presidential Address to the International Society for Clinical Enzymology, London, 19 September 1977.

When rat lymphocytes are incubated at 37°C in Krebs-Ringer solution, there is little leakage of lactate dehydrogenase into the medium during the first 2 h, during which time the intracellular ATP gradually falls to less than 20% of its original concentration. Subsequently, enzyme efflux occurs at a much faster rate [24].

Other high-energy phosphates, e.g. phosphoenolpyruvate, also reduce enzyme leakage. Glucose exerts a slight protective effect, but this is reversed in the presence of inhibitors of glycolysis, e.g. iodoacetate, or uncouplers of oxidative phosphorylation, e.g. 2,4-dinitrophenol [7].

It seems therefore that their energy content is an important factor protecting cells against enzyme release, and its reduction is likely to follow pathological processes involving interference with cellular oxygenation, nutrition, etc.

The diminution in the rate of intracellular enzyme release reported in the experimental studies cited above occurred when ATP was incorporated into the external medium. The cell membrane, however, is considered to be impermeable to external ATP and, if this is correct, then the protective effect must be exerted at the external surface of the cell membrane, where membrane-bound ATPase may be able to utilize extracellular ATP, thus conserving intracellular energy supplies.

The intracellular energy content, however, does not appear to be the only factor involved in maintaining the integrity of the cell membrane since AMP has been found to reduce enzyme release from human lymphocytes, in marked contrast to its effect on rat lymphocytes where it causes an increase in enzyme efflux [9]. Attempts to explain this species difference on the possible occurrence of an enzyme system capable of converting AMP into ATP in the human cells have proved inconclusive, and it appears that AMP may exert a direct effect on the permeability of the human cell membrane.

Also difficult to reconcile with the hypothesis that the intracellular energy content controls the permeability of the cell membrane to intracellular enzymes is the observation that lymphocytes from patients with chronic lymphocytic leukaemia lose intracellular enzymes during prolonged incubation at slower rates than normal human lymphocytes despite their significantly lower ATP contents [8]. Such leukaemic lymphocytes, however, have been shown to possess an outer coating [13], which may be responsible for the reduction in the rate of enzyme efflux. Nevertheless, incorporation of ATP into the medium reduces enzyme leakage from both leukaemic and normal lymphocytes.

The Disappearance of Enzymes from the Plasma

Though there is much clinical evidence showing that enzymes disappear from the plasma at different rates, the mechanisms involved have remained obscure. Experimental studies have suggested that the reticulo-endothelial system [21] or the small intestine [6] may be concerned. The demonstration that experimental removal of the liver, spleen or kidneys does not affect the disappearance rates of injected enzymes [18] appears to exclude excretory processes, and it is well-known that in the absence of renal disease no more than traces of enzymes appear in the urine, apart from those of low molecular mass, e.g. amylase, uropepsinogen, which can be filtered by the glomerulus. It has therefore been suggested that circulating enzymes might undergo intravascular inactivation [14, 22], but until recently there has been no evidence to support this theory. KATUNUMA *et al.* [10] have reported that rat tissues contain enzymes specifically hydrolysing circulating enzymes, but this awaits confirmation.

A number of attempts have been made to study enzyme catabolism with radioactively labelled enzymes on the assumption that the label would enable the presence of enzyme protein to be detected even after it had lost its catalytic activity. In all experiments with ^{131}I-labelled enzymes [12, 17] or with enzymes coupled with ^{14}C-labelled iodoacetate [3] or acetic anhydride [2], the radioactivity was found to disappear from the plasma at the same rate as the catalytic activity. This suggests that the enzyme protein is removed without previous inactivation. However, the energy of γ-emission from ^{131}I has been shown to cause enzyme inactivation [2], and there is evidence that enzyme conjugates may be rapidly hydrolysed *in vivo* [15]. Lactate dehydrogenase-5 labelled by coupling with a 3,5-diiodo-4-hydroxypropionyl ester has been shown to break down in the rabbit at such a rate that the bulk of the label appears in the urine in a few hours [15]. In animals injected with such a labelled enzyme, the circulating radioactivity is probably associated largely with the hydrolysed coupling agent rather than with the enzyme protein.

Dr. QURESHI and I [16, 23] therefore decided to label rabbit lactate dehydrogenases-1 and -5 with ^{125}I since the energy of γ-emission from this isotope is much lower than that from ^{131}I and would be less likely to inactivate the enzyme. Furthermore, its much longer half-life allows more time for purification of the labelled enzyme and enables the stability of the latter to be established. Since radioiodine atoms are covalently bound to tyrosyl or histidyl groups in the enzyme molecule, there is little risk of the

Table I. Biological half-lives for unlabelled and ^{125}I-labelled lactate dehydrogenase isoenzymes 1 and 5 in the plasma of the rabbit

Isoenzyme	Phase	$t_{0.5}$, h, measured by[1]		
		unlabelled enzyme	labelled enzyme	
		catalytic activity	catalytic activity	radioactivity
1	1	0.53, 0.52[3]	0.62±0.07	0.41±0.13 [2]
				1.39±0.31 [2]
	2	17.33, 17.77[3]	18.36±0.45	30.54±1.73
	3	–	–	80[3]
5	1	0.66±0.10	0.70±0.24	0.33±0.10 [2]
				0.83±0.31 [2]
	2	3.48±0.28	3.50±0.50	12.41±2.30
	3	–	–	50±6.2

[1] Means of 4 or 5 experiments ±SD.
[2] The curves representing the disappearance of radioactivity during phase 1 were resolved into two exponential components of unknown significance.
[3] Results of single experiments.

label being removed from the enzyme soon after intravenous injection into rabbits.

The animals were given sodium ^{127}iodide in their drinking water to saturate the thyroids and the following day were injected with one or other of the labelled enzymes. Blood samples were collected at frequent intervals for measurement of the plasma radioactivity and catalytic activity. The resulting curves were plotted on semi-log paper and analysed by a curve-splitting technique [4]. A three-phase exponential system was obtained. The first phase (1–4 h after injection of ^{125}I-LD-5 and up to 8 h after ^{125}I-LD-1) is attributed to the distribution of the injected material throughout the extracellular fluid. The second phase continued until the catalytic activity became constant, after which the residual radioactivity disappeared at a much slower rate.

The biological half-lives for the disappearance of the two enzymes in each of the three phases are summarized in table I. Radioactivity and catalytic activity disappeared at similar fast rates during phase 1, but during phase 2 catalytic activity disappeared from the plasma at significantly faster rates than the radioactivity ($p<0.001$ in each case), but at similar rates to the unlabelled enzyme. This suggests that the presence of the radioactive label does not affect the manner in which the enzyme is catabolized.

The faster disappearance of catalytic activity during the catabolic phase 2 indicates that the injected enzyme is inactivated either in the circulating blood or in a closely associated compartment, e.g. the spleen, since catalytically inactive, but radioactive, enzyme protein can still be detected in the plasma during this phase.

No tissue was found to contain greater radioactivity than the plasma; hence, it seems that removal of the inactivated enzyme protein or its breakdown products is a passive process. The intestinal mucose and intestinal contents exhibited appreciable radioactivity (20–50% of the plasma radioactivity) 2–8 h after injection, but little appeared in the faeces. About 60–70% was excreted in the urine within 72 h of injection. Two thirds was identified as mono- and di-iodotyrosines, di-iodohistidine or their metabolites, while the remainder consisted of ^{125}iodide.

We conclude therefore that lactate dehydrogenase isoenzymes-1 and -5 are metabolized by similar mechanisms, though of course at different rates. It seems that the enzymes are inactivated in the circulating blood and that the inactivated enzyme proteins or their breakdown products are discharged into the intestine either directly from the mucosal cells or indirectly via the liver and the bile. In the intestine, further digestion occurs, and the resulting amino acids or small peptides are reabsorbed to enter the amino acid pool. Iodotyrosines and iodohistidine, however, cannot be utilized in the biosynthesis of protein and in the experiments described these amino acids appeared in the urine. The presence of appreciable amounts of iodide is probably due to hepatic deiodination of iodotyrosines [11].

The proposed mechanism is similar to that suggested for albumin by Armstrong *et al.* [1] and since confirmed by a variety of techniques. It remains to be established whether such mechanisms apply to the catabolism of other enzymes and whether man gets rid of surplus enzymes in this way.

References

1 Armstrong, F.B.; Margen, S., and Tarver, H.: Plasma protein VII. Site of degradation of serum albumin. Proc. Soc. exp. Biol. Med. *103:* 592–596 (1960).

2 Bär, U.; Friedel, R.; Heine, H.; Mayer, D.; Ohlendorf, S.; Schmidt, F.W., and Trautschold, I.: Studies on enzyme elimination. III. Distribution, transport and elimination of cell enzymes in the extracellular space. Enzyme *14:* 133–156 (1972/1973).

3 Boyd, J.W.: The disappearance of ^{14}C-labelled isoenzyme-5 of L-lactate dehydrogenase from plasma. Biochim. biophys. Acta *146:* 590–593 (1967).

4 Donato, L.; Matthews, C.M.E.; Nosslin, B.; Segre, G., and Vitek, F.: Round

table on applications of tracer theory to protein turnover studies. J. Nucl. Biol. Med. *10:* 3–30 (1966).

5 ENGLHARDT, A.; SCHMIDT-SODINGEN, G. und LANGE, H.: Metabolitgehalt und Enzympermeabilität isolierter menschlicher Blutzellen bei Substratmangel und Zusatz von Stoffwechselgiften. Enzymol. biol. clin. *10:* 258–280 (1969).

6 FLEISHER, G.A. and WAKIM, K.G.: The role of the small intestine in the disposal of enzymes. Enzymol. biol. clin. *9:* 81–96 (1968).

7 HALLAK, G.J. and WILKINSON, J.H.: Action of metabolic inhibitors on the release of intracellular enzymes from human and rat lymphocytes and human erythrocytes. Clin. chim. Acta *66:* 251–261 (1976).

8 HALLAK, G.J. and WILKINSON, J.H.: Membrane permeability in normal human lymphocytes and lymphocytes from patients with chronic lymphatic leukaemia. Clin. chim. Acta *69:* 341–349 (1976).

9 HALLAK, G.J. and WILKINSON, J.H.: Action of adenosine phosphates on the release of intracellular lactate dehydrogenase from human and rat lymphocytes. Enzyme *22:* 361–369 (1977).

10 KATUNUMA, N.; KOMINAMI, E.; KOMINAMI, S., *et al.:* Regulatory mechanism of enzyme catabolism; in WIELAND, HELMREICH and HOLZER 2nd Int. Symp. on Metabolic Interconversion of Enzymes, 1971, p. 159 (Springer, Berlin 1972).

11 MACLAGAN, N.F. and REID, D.: The deiodination of thyroid hormones *in vitro;* in WOLSTENHOLME and MILLAR 10th Ciba Found. Coll. on Endocrinology. Regulation and mode of action of thyroid hormones, pp. 190–203 (Churchill, London 1957).

12 MASSARAT, S.: Enzyme kinetics, half-life, and immunological properties of iodine- 131-labelled transaminases in pig blood. Nature, Lond. *206:* 508–509 (1965).

13 PERERA, D.J.B. and PEGRUM, G.D.: *In vitro* stimulation of chronic lymphocytic leukaemic lymphocytes. Br. J. Haemat. *27:* 289–301 (1974).

14 POSEN, S.: Turnover of circulating enzymes. Clin. Chem. *16:* 71–84 (1970).

15 QURESHI, A.R. and WILKINSON, J.H.: Rates of disappearance from plasma of enzymes labeled by coupling with a radioactive iodo-ester. Clin. Chem. *22:* 1277–1282 (1976).

16 QURESHI, A.R. and WILKINSON, J.H.: The fate of circulating lactate dehydrogenase-5 in the rabbit. Clin. Sci. mol. Med. *50:* 1–14 (1976).

17 SCHAPIRA, F.; DREYFUS, J.-C. et SCHAPIRA, G.: La durée de séjour dans le plasma de l'aldolase chez le lapin: étude à l'aide d'une aldolase marquée à l'iode radioactif. Rev. fr. Etud. clin. biol. *7:* 829–832 (1962).

18 STRANDJORD, P.E.; THOMAS, K.E., and WHITE, L.P.: Studies on isocitric and lactic dehydrogenase in experimental myocardial infarction. J. clin. Invest. *38:* 2111–2118 (1959).

19 SWEETIN, J.C. and THOMSON, W.H.S.: Enzyme efflux and clearance. Clin. chim. Acta *48:* 403–411 (1973).

20 TATO, L. and RUBALTELLI, F.F.: The protective action of glucose and nucleotides against haemolysis induced by menadione and phospholipase A. Clin. chim. Acta *49:* 349–352 (1973).

21 WAKIM, K.G. and FLEISHER, K.G.: The fate of enzymes in body fluids – an experimental study. II. Disappearance rates of glutamic-oxalacetic transaminase I under various conditions. J. Lab. clin. Med. *61:* 86–97 (1963).

22 Wilkinson, J.H.: Clinical significance of enzyme activity measurements. Clin. Chem. *16:* 882–890 (1970).

23 Wilkinson, J.H. and Qureshi, A.R.: Catabolism of plasma enzymes, as studied with ^{125}I-labelled lactate dehydrogenase-1 in the rabbit. Clin. Chem. *22:* 1269–1276 (1976).

24 Wilkinson, J.H. and Robinson, J.M.: Effect of energy-rich compounds on release of intercellular enzymes from human leukocytes and rat lymphocytes. Clin. Chem. *20:* 1331–1336 (1974).

25 Wilkinson, J.H.; Robinson, J.M., and Johnson, K.P.: Factors affecting the release of haemoglobin and enzymes from human erythrocytes. Ann. clin. Biochem. *12:* 58–65 (1975).

26 Zierler, K.L.: Diffusion of aldolase from rat skeletal muscle. An index of membrane permeability. Am. J. Physiol. *190:* 201–205 (1958).

J.H. Wilkinson, DSc, PhD, Department of Chemical Pathology, Charing Cross Hospital Medical School, *London* (England)

Enzymes in Health and Disease. Inaug. Scient. Meet. Int. Soc. Clin. Enzymol., London 1977, pp. 8–18 (Karger, Basel 1978)

Pathophysiology of Enzymes

JEAN-CLAUDE DREYFUS
Institut de Pathologie Moléculaire, Paris

Enzyme dysfunction and inactivation represent one of the most important chapters in the field of biochemical pathology. Their mechanisms, however, are still largely unknown, and subject to hypotheses which rest mainly on what has been learnt from the study of abnormal hemoglobins. In this review, we shall attempt to define the major probable mechanisms of the physiopathology of enzymes, giving examples whenever possible.

I. General Mechanisms

Alterations of enzymes can result from mutational or non-mutational events (table I).

(A) Mutational events are in most cases due to the mutation of a single base in DNA, causing the replacement of one amino acid by another, and resulting in a missense mutation. We shall discuss them more in detail below.

(B) Non-mutational events are due to various mechanisms. Table I summarizes some of these events. All known examples of these mechanisms come from the field of abnormal hemoglobins.

The case for controller or regulatory gene disease deserves some comments: following the work on bacteria by JACOB and MONOD [11], the notion of controller gene disease has been widely applied [22]. In earlier papers, we have attempted to define the criteria which should be fulfilled for controller genes of the bacterial type (e.g. diffusible proteins possessing an affinity for a defined part of the genome) to apply to human diseases (table II) [3, 4]. Judging by these criteria, very few diseases remained good candidates for being controller gene diseases. The best seemed to be thalassemias

Table I. General mechanisms of protein abnormalities

A. Mutational events
1. Missense mutation → replacement of an amino acid by another one
2. Nonsense mutation (stop signal) → no synthesis or synthesis of an incomplete protein
3. Frame-shift mutation → synthesis of a completely abnormal protein
4. Lack of a terminating codon → increased molecular weight

B. Non-mutational events
Unequal crossing over
Non-functional or unstable messenger RNA
Deletion of one or several genes, complete or incomplete
Abnormality of a regulatory gene

Table II. Theoretical aspects of controller gene mutations

Control system	Type	Function	Expression
Negative	regulator negative	gain	recessive
Negative	superrepressor	loss	dominant
Positive	regulator negative	loss	recessive
Positive	constitutive	gain	dominant

and some porphyrias, the major biochemical feature of which is an increase in production of δ-aminolevulinic acid.

A few years later, none of these candidates could stand new accumulated evidence. Although their mechanisms are multiple, none of the thalassemias fulfills the criteria of a controller gene disease, and most porphyrias have been shown to be due to specific enzyme deficiencies. Some new candidates, however, arose, and two examples may be given. One is the case of polyenzymatic diseases (see below). Another is the deficiency of phosphorylase kinase. This disease exists in man, but has been better studied in the I strain of mice, displaying a sex-linked recessive deficiency of phosphorylase kinase in muscle. Features include the following: deficiency is complete in muscle, incomplete in heart, while it has not been possible to differentiate muscle and heart enzyme [16]; presence of enzyme activity in the muscle during the first days or weeks of life [16]; no decrease or only slight decrease of activity of muscle in heterozygotes [9], and absence of the molecule in deficient muscle [2].

Table III. Recessive and dominant diseases

A. Recessive diseases: deficiency in a biological activity
Enzymic deficiencies
Hemophilia
Afibrinogenemia
B. Dominant diseases: deleterious effects of the abnormal product
Dysfibrinogenemia
Unstable hemoglobins
Partial deficiency in a limitant enzyme
Several types of porphyrias

A. Recessive and Dominant Diseases

For a long time, only the effects of recessive diseases could be interpreted in molecular terms. Dominant diseases are more difficult to understand, since the normal allele is producing the normal product. Pleiotropic dominant diseases are still beyond our understanding, but in many cases a molecular mechanism is now at hand. Table III summarizes our present knowledge.

II. Effects of Missense Mutations

Missense mutations may have various consequences on the activity of enzymes. Alterations due to mutations may be: (a) none at all; (b) loss or decrease in activity; (c) decrease in stability; (d) decrease in both activity and stability; (e) abnormal functional features; (f) impossibility of transformation of an inactive proenzyme into an active enzyme.

These various possibilities will be illustrated for the four first categories by taking examples of anomalies of glucose-6-phosphate dehydrogenase, the enzyme for which the greatest number of variants is known (see [15] for review).

A. No Decrease in Activity

Mutants with a normal or nearly normal activity are known. Some of them may be easily recognized by their modified electrophoretic mobility. A very common variant among Negroes, G6PD A^+, migrates faster towards the anode than the wild allele B^+. By contrast, other variants with nearly normal activity migrate more slowly. A very striking case is that of G6PD Hektoen, marked by a substitution of a tyrosine for a histidine, and whose rate of synthesis is much faster than normal [28]. This proves that a single mutational event may influence not only the function and stability of a protein, but the regulation of its synthesis as well.

B. Decrease in Stability

This decrease in stability has consequences which depend on the severity of the decrease.

(1) A moderate decrease in stability only affects cells with a very slow or no protein synthesis. The deficiency is only observed in red blood cells, and especially in the old cells. The ratio of activity old/young RBC is lower than normal. White blood cells have a normal activity. The most typical example is the variant G6PD A^-, very frequent in Negro populations.

(2) A severe decrease in stability ends in a low activity in all tissues, including white blood cells, and fibroblasts in culture. An example of this possibility is the Benevento-like variant [12].

C. Decrease in Activity with a Normal Stability

The effect of this type of mutation is to decrease the specific activity of the molecule, e.g. the ratio of the activity of the enzyme to the number of molecules immunologically determined. Immunological determinations of enzymes can be made by several techniques: immunoneutralization, antibody consumption, complement deviation, immunodiffusion in agar with or without electrophoresis, and radioimmunology.

Few examples of such types are known. A few years ago, we described G6PD Port Royal, with an activity of 40% of normal and no decrease of the number of molecules [25].

D. Decrease in Both Stability and Efficiency

These mutations decrease the molecular specific activity of the enzyme and also the activity. Several such variants are known, and they are probably many [29], but for most variants the immunological test has not yet been made. The best known example is G6PD B^-, also called Mediterranean variant, whose frequency is very great. The activity in red blood cells is 0–5% of normal. In white cells it is 30% of normal but, since the specific activity of the molecule is also decreased to 30% of normal, the number of molecules is the same as that of the controls in WBC, and decreased only in RBC [12, 25, 29].

E. Abnormal Functional Features of Pathological Enzymes

In several enzymatic diseases, the activity was first considered as being absent. Later, it was recognized that alterations of the molecule led to the apparent inactivity, but that it was possible to detect some activity by modifying the conditions of the assay.

(1) The affinity for the substrate can be greatly decreased. An increase in the concentration of substrate will reveal the residual activity [4].

(2) The affinity or binding properties towards the coenzyme can be altered. Known examples are several enzymes requiring pyridoxal phosphate or vitamin B_{12} derivatives [4]. This case is of course of special interest, since it allows us to obtain a therapeutic effect by giving the patient high doses of the relevant vitamin or coenzyme.

III. Presence or Absence of the Molecule

Immunological techniques are capable of revealing the presence of the molecule of inactive enzymes in many cases. This presence is the witness of an altered molecule and, therefore, proves a structural mutation. A good example of the persistence of a molecule, demonstrated by the presence of cross-reacting material (CRM), is the disease hereditary fructose intolerance [21], whose biochemical defect is a decreased activity of the specific liver

Table IV. Presence or absence of cross-reacting immunological material (CRM) in lysosomal diseases, indicated by name of deficient enzymes

Enzyme	CRM	Disease
β-Hexosaminidase	−	Tay-Sachs and Sandhoff diseases, gangliosidosis GM2
α-Galactosidase	−	Fabry's disease
β-Galactosidase	+	Landing's disease, gangliosidosis GM1
α-Mannosidase	+	mannosidosis
α-Fucosidase	−	fucosidosis
Arylsulfatase A	+	metachromatic leukodystrophy

Cells tested were usually cultured fibroblasts or leukocytes. These lysosomal enzymes are present in all tissues.

aldolase (aldolase B). No clue concerning the presence or absence of this CRM can be deduced from the nature of the involved enzyme or of the disease. Some deficiencies of the same enzyme can encompass both types, with and without the molecule. As an example, we give in table IV a short list of lysosomal diseases. It can be seen that enzymes as close as α- and β-galactosidases show respectively the absence or presence of CRM.

IV. Age of Onset and Severity of the Disease

When enzymatic diseases began to be known, it was first assumed that, given one altered enzyme, the metabolic and clinical consequences were to be the same in all affected patients. This rule has now been disproved in many cases. One evident cause of clinical polymorphism is incomplete deficiencies like those of glucose-6-phosphate dehydrogenase. Even when the deficiency is apparently complete, discrepancies are found. The most striking case is the deficiency of α-L-iduronidase, observed both in the very severe Hurler's disease (mucopolysaccharidosis type I) and the mild Scheie's disease (mucopolysaccharidosis type V) [27]. The provisional explanation, of two different alterations of the same gene, is likely to be correct, but has to be confirmed in order to really explain the facts.

Another interesting observation is that of the existence of early and late forms of the same disease. The deficiency of several lysosomal enzymes, α-glucosidase, β-galactosidase, β-glucosidase, β-hexosaminidases, and others, are examples of this phenomenon. A very severe disease in the first months of life is the typical feature, leading to death after a few months or at most one or 2 years of life. By contrast, in other patients the symptoms appear only in late childhood or even after puberty, and the course is milder, permitting in some cases a normal life span. One good example is the deficiency of amylo-1,4-glucosidase or acid maltase. The infant form is the well-known Pompe's disease, lethal after about 1 year because of cardiac failure. Contrasting with that picture is the late form of the disease, which is recognized at any age, and even sometimes in elderly persons, and whose very polymorphic clinical picture is generally dominated by myopathic symptoms. It was generally accepted that the deficiency was as complete in the latter form as in the former one. In 1972, we described the presence of a residual activity in a boy of 8 [5]. It has now been recognized that with sensitive methods residual activity of the enzyme can always be detected [17].

Similarly, the presence of a small residual activity in late onset disease has also been demonstrated in other deficiencies such as that of β-galactosidase.

In all types of enzymatic diseases, the genetic form is constant in a given family, and there is no coexistence, for example, of Pompe's disease and the myopathic form in the same family.

Recently, one exception has been observed: early and late types of fucosidosis, a disease due to the lack of α-fucosidase, have been described. The presence of both types has been demonstrated in the same family in two instances [24]. This casts doubt on the genetic origin of these two clinical types of fucosidosis.

V. Polyenzymatic Diseases

This section encompasses diseases which can involve more than one enzyme protein. This involvement must always be considered with caution, since an apparent lack of activity might be due to improper handling of a biological sample. Incontestable cases, therefore, are rare.

(1) The most conclusive demonstration of the lack of two enzymatic activities is orotic aciduria [10]. In this disease, two consecutive enzymes of the pyrimidine biosynthetic pathway are lacking, orotidylic pyrophosphorylase and decarboxylase. Two types of mechanisms can be proposed: (a) a regulatory gene, and (b) the existence of a common subunit.

(2) No definitive conclusion can be drawn as regards orotic aciduria. The existence of common subunits, however, has been put forward in several enzymatic diseases, and proved in one type of deficiency, that of β-hexosaminidases. It is now generally accepted that Hex B is a homopolymer of β-subunits, while Hex A is a heteropolymer of α and β subunits; the β subunit, therefore, is shared by both isoenzymes [1]. Its absence suppresses the activity of both isoenzymes A and B (Sandhoff's disease), the absence of α subunit leaves Hex B intact but results in the disappearance of Hex A (Tay-Sachs disease).

(3) A third type of polyenzymatic disease is examplified by the so-called mucolipidoses type II (I cell disease) and type III [20]. Many lysosomal enzymes are found to be decreased in cultured fibroblasts, and greatly increased in plasma. No definitive explanation can be given yet, but it seems likely that a common abnormality exists on the biosynthetic pathway, bearing perhaps on the carbohydrate moiety of these enzymes.

VI. Behaviour in vivo *and* in vitro

One of the greatest advances of the last years has been the demonstration that a deficiency found in a patient can be also observed in fibroblasts obtained from the patient's skin and cultured *in vitro*, provided the deficient enzyme is present in normal fibroblasts. This discovery has allowed us to go more deeply into the pathogenesis of many diseases and, most of all, to establish a firm basis for the antenatal diagnosis of hereditary metabolic diseases. Exceptions, however, may be found. We observed that cultured human fibroblasts express the proteins coded by two genetic types of glycogen phosphorylase, e.g. the liver type and the brain type [23]. It appears that in patients having a true deficiency in liver phosphorylase, the fibroblasts still express liver phosphorylase activity like the controls [13]. An analogous observation was made with muscle phosphorylase deficiency. Fibroblasts do not possess muscle phosphorylase activity. But it is possible to culture muscle cells, which keep muscle phosphorylase expression, mixed with brain phosphorylase activity. In two patients of this type, we were able to detect muscle phosphorylase activity in the culture [18].

VII. Pseudodeficiencies

One well-known danger in medical enzymology is to overlook the existence of a deficiency because for convenience we use artificial substrates and not the natural substrates. The reverse may also happen.

During screening programs – or when looking for heterozygotes in families at risk for a deficiency – apparent complete or nearly complete absence of activity may be found in healthy persons. Two such cases have been well documented.

(1) The deficiency of arylsulfatase A is responsible for the disease metachromatic leukodystrophy. In two families at risk, siblings and one parent were found apparently as deficient as the sick proband [8, 14]. The problem was made even more difficult by the fact that the leukocytes of these pseudo-deficient siblings were also inactive against the physiological substrate cerebroside-sulfate. An antenatal diagnosis appears impossible in such families.

(2) The case for hexosaminidases is still more complicated. In two families, a deficiency of Hex A was found with no symptoms of Tay-Sachs disease [19, 26]. Later determinations with more sensitive techniques showed

that a small residual activity was actually present, contrary to what is found in genuine Tay-Sachs patients.

We [6, 7] have found a family at risk in which the father showed a very low activity of total hexosaminidase, mimicking Sandhoff's disease, characterized by the absence of both Hex A and B. Cellulose acetate electrophoresis showed the presence of a faint band of Hex A, amounting to about 10% of normal, with no visible B band. When his wife became pregnant, cells from the amniotic fluid were cultured. Electrophoresis of the extract from these cells showed the same picture as that of the father. The pregnancy was allowed to continue, and the child was clinically normal at the age of 16 months.

These observations pose difficult problems in the field of prenatal diagnosis. They show also that a normal life can be led with activities of vital enzymes no greater than 10% of normal, e.g. 20% of the value of heterozygotes, and pose the question, 'What is the lowest limit compatible with normal life'.

Summary

An overall view of the mechanisms of enzyme abnormality is presented. It includes: (1) general mechanisms (mutation vs controller genes); (2) discussion of the effects of missense mutations, taking as examples variants of glucose-6-phosphate dehydrogenase; (3) presence or absence of the molecule; (4) age of onset and severity of the disease; (5) polyenzymatic diseases; (6) pseudodeficiencies.

References

1 BEUTLER, E. and KUHL, W.: Subunit structure of human hexosaminidase verified; interconvertibility of hexosaminidase isozymes. Nature, Lond. *258:* 262–264 (1975).

2 COHEN, P.T.W.; BURCHELL, A., and COHEN, P.: The molecular basis of skeletal muscle phosphorylase kinase deficiency. Eur. J. Biochem. *66:* 347–356 (1976).

3 DREYFUS, J.C.: The application of bacterial genetics to the study of human genetic abnormalities; in STEINBERG and BEARN Progr. med. Genet., vol. 6, pp. 169–200 (Grune & Stratton, New York 1969).

4 DREYFUS, J.C.: Bases moléculaires des maladies enzymatiques génétiques. Biochimie *54:* 559–571 (1972).

5 DREYFUS, J.C. and ALEXANDRE, Y.: Electrophoretic characterization of acidic and neutral amylo 1–4 glucosidase (acid maltase) in human tissues. Biochem. biophys. Res. Commun. *48:* 914–920 (1972).

6 DREYFUS, J.C.; POENARU, L., and SVENNERHOLM, L.: Absence of hexosaminidases A and B in a normal adult. New Engl. J. Med. *292:* 61–63 (1975).
7 DREYFUS, J.C.; POENARU, L.; VIBERT, M.; RAVISE, N., and BOUÉ, J.: Characterization of a variant of β-hexosaminidase: 'Hexosaminidase Paris'. Am. J. hum. Genet. *29:* 287–293 (1977).
8 DUBOIS, G.; TURPIN, J.C., and BAUMANN, N.: Metachromatic leukodystrophy and arylsulphatase A: relations and discrepancies. Biomedicine *26:* 317–319 (1977).
9 GROSS, S.L.; LONGSHORE, M.A., and PANGBURN, S.: The phosphorylase kinase deficiency locus in the mouse: evidence that the mutant allele codes for an enzyme with an abnormal structure. Biochem. Genet. *13:* 567–584 (1975).
10 HUGULEY, C.M., jr.; BAIN, J.A.; RIVERS, S.L., and SCROGGINS, R.B.: Refractory megaloblastic anemia associated with excretion of orotic acid. Blood *14:* 615–634 (1959).
11 JACOB, F. and MONOD, J.: Genetic regulatory mechanisms in the synthesis of proteins. J. molec. Biol. *3:* 318–356 (1961).
12 KAHN, A.: G6PD variants. Hum. Genet. (in press).
13 KOSTER, J.F.; SLEE, R.G.; DAEGELEN, D.; MEIENHOFER, M.C.; DREYFUS, J.C.; NIERMEYER, M.F., and FERNANDES, J.: Isoenzyme pattern of phosphorylase in white blood cells and fibroblasts from patients with liver phosphorylase deficiency. Clin. chim. Acta *69:* 121–125 (1976).
14 LOTT, I.T.; DUNALEN, J.T.; MILUNSKI, A.; HOEFNAGEL, D., and MOSER, H.W.: Apparent biochemical homozygote in two obligate heterozygotes for metachromatic leukodystrophy. J. Pediat. *89:* 438–440 (1976).
15 LUZZATTO, L.: Genetic heterogeneity and pathophysiology of G6PD deficiency. Br. J. Hematol. *28:* 151–155 (1974).
16 LYON, J.B., jr.: The X chromosome and the enzymes controlling muscle glycogen phosphorylase kinase. Biochem. Genet. *4:* 169–185 (1970).
17 MEHLER, M. and DIMAURO, S.: Residual acid maltase activity in late-onset acid maltase deficiency. Neurology, Minneap. *27:* 178–184 (1977).
18 MEIENHOFER, M.C.; ASKANAS, V.; DAEGELEN, D.; DREYFUS, J.C., and ENGEL, W.K.: Muscle-type phosphorylase activity present in muscle cells cultured from three patients with myophosphorylase deficiency. Archs Neurol., Chicago *34:* 779–781 (1977).
19 NAVON, R.; PADEH, B., and ADAM, A.: Apparent deficiency of hexosaminidase A in healthy members of a family with Tay-Sachs disease. Am. J. hum. Genet. *25:* 287–293 (1973).
20 NEUFELD, E.F.: The biochemical basis for mucopolysaccharidoses and mucolipidoses; in STEINBERG and BEARN Prog. med. Genet., vol. 10, pp. 81–101 (Grune & Stratton, New York 1974).
21 NORDMANN, Y.; SCHAPIRA, F., and DREYFUS, J.C.: A structurally modified liver aldolase in fructose intolerance: immunological and kinetic evidence. Biochem. biophys. Res. Commun. *31:* 884–889 (1968).
22 PARKER, W.C. and BEARN, A.G.: Application of genetic regulatory mechanisms to human genetics. Am. J. Med. *34:* 680–691 (1963).
23 PROUX, D.; VIBERT, M.; MEIENHOFER, M.C., and DREYFUS, J.C.: The isozymes of glycogen phosphorylase in human and rabbit tissues. Electrofocusing in polyacrylamide gels. Clin. chim. Acta *57:* 211–216 (1974).

24 ROMEO, G.; BORRONE, G.; GATTI, R., and DURAND, P : Fucosidosis in Calabria: founder effect or high gene frequency? Lancet *i:* 368–369 (1977).
25 ROSA, R.; ALEXANDRE, Y.; KAPLAN, J.C. et DREYFUS, J.C.: Comportement immunologique de la glucose-6-phosphate deshydrogénase érythrocytaire chez des mutants. Clin. chim. Acta *29:* 209–214 (1970).
26 VIDGOFF, J.; BUIST, N.R.M., and O'BRIEN, J.S.: Absence of β N-acetyl D-hexosaminidase A activity in a healthy woman. Am. J. hum. Genet. *25:* 372–381 (1973).
27 WIESMANN, U. and NEUFELD, E.F.: Scheie and Hurler syndromes: apparent identity of the biochemical defect. Science *169:* 72–74 (1970).
28 YOSHIDA, A.: Aminoacid substitution (histidine to tyrosine) in a glucose-6-phosphate dehydrogenase variant (G6PD Hektoen) associated with overproduction. J. molec. Biol. *52:* 483–490 (1970).
29 YOSHIDA, A.; STAMATOYANNOPOULOS, G., and MOTULSKY, A.G.: Biochemical genetics of glucose-6-phosphate dehydrogenase variation. Ann. N.Y. Acad. Sci. *155:* 868–877 (1968).

J.-C. DREYFUS, MD, Institut de Pathologie Moléculaire, 24, rue du Faubourg Saint-Jacques, *F-75014 Paris* (France)

Enzymes in Health and Disease. Inaug. Scient. Meet. Int. Soc. Clin. Enzymol., London 1977, pp. 19–22 (Karger, Basel 1978)

Platelets and the Release of Lactate Dehydrogenase

Renze Bais and John B. Edwards

Division of Clinical Chemistry, Institute of Medical and Veterinary Science, Adelaide

The release of lactate dehydrogenase (LD) from platelets can be achieved by the use of detergents [1] or by lowering the osmolality [5]. These phenomena result in a lysis of the platelets and the release of the enzyme. However, a more subtle release of LD may be achieved by heating platelets in the presence of heparin. This initial observation has been investigated further, after it was demonstrated that differentiating isoenzymes of LD by assaying the enzyme before and after heating for 15 min at 60 °C resulted in the activity after heating being greater than the initial value. These experiments were carried out on plasma that had been prepared by centrifuging the heparinized blood at 1,800 *g* for 5 min. Under these conditions, not all the platelets were removed and they were able to contribute to the total LD activity. In order to fully understand this finding, plasma formed in the presence of the anticoagulants, heparin (12.5 U/ml), citrate (0.106 M) or EDTA (1–2 mg/ml) was compared with serum. The procedure involved heating the sample at 60 °C, cooling rapidly, centrifuging and assaying the resultant supernatant. The results (fig. 1) showed that heparinized plasma gave a large and rapid increase in LD activity whereas EDTA and citrate behave in a manner similar to serum. Electrophoresis of the heated plasma indicated an increase in isoenzymes 2 and 3 which is consistent with an involvement of platelets.

Further evidence that platelets were the cause of the increase in LD activity was obtained using labelled serotonin. Platelets readily take up serotonin even though the uptake occurs against a concentration gradient. The process requires energy and is temperature-dependent [6]. In our experiments, heparinized blood was allowed to equilibrate with ^{3}H-serotonin for 15 min at 37 °C. The plasma was then heated at 50, 55 and 60 °C for various times, an aliquot removed, centrifuged and the enzymic activity and

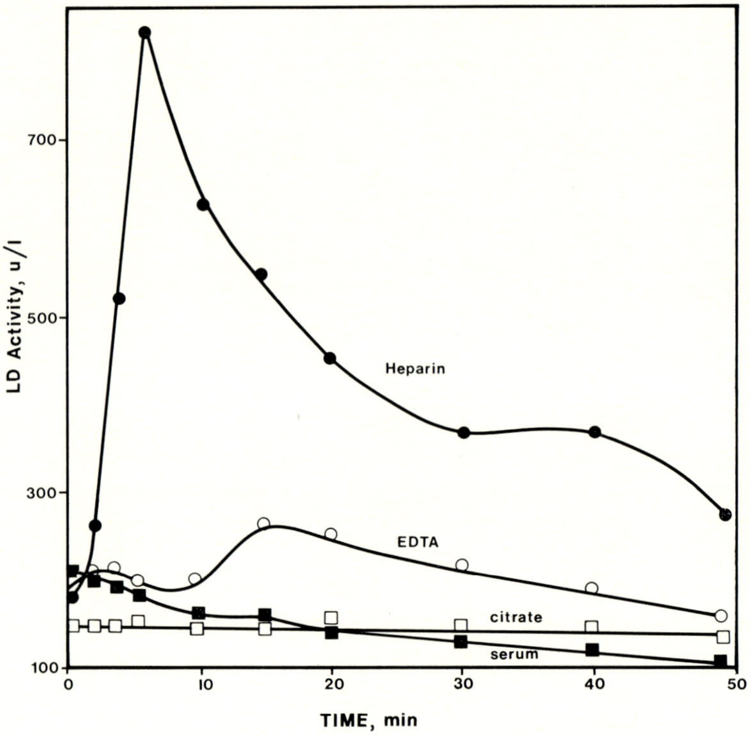

Fig. 1. A comparison of the release at 56 °C of lactate dehydrogenase from serum and plasma formed in the presence of the anticoagulants, heparin (12.5 U/ml), citrate (0.106 M) or EDTA (1–2 mg/ml). After the initial rapid increase in LD activity, due to enzyme release as described in the text, the decrease in activity is due to destruction of the enzymic activity at 60 °C. Lactate dehydrogenase activity was determined on a Beckman Enzyme Activity Analyzer System TR by the method of WACKER *et al.* [7].

radioactivity of the supernatant determined. The results show that there is a parallel release of enzymic activity and labelled serotonin, implicating platelets as the source of the LD. In addition, the parallel release of LD and serotonin only occurs with heparinized plasma; when EDTA is used, serotonin is released but the LD is not. These results indicate that there may be a two-phase release process operating from platelets, i.e. heating in the presence of EDTA causes a change in the platelet membrane and allows leakage of small molecules, whereas heating in the presence of heparin may result in lysis of the platelets and release of all the cytoplasmic constituents. The differential rate of release for various substances has been discussed by

HOLMSEN *et al.* [4], who suggested that the release reaction could also be induced by a variety of stimuli including collagen [3]. In our experiments, platelets incubated with collagen show a slow release of enzymic activity and do not mimic the heat release process.

These results using anticoagulants indicate that the release of enzyme only occurs with heparinized plasma. Furthermore, the addition of EDTA or citrate to heparinized plasma prevents the release of enzyme, but the addition of heparin to EDTA or citrated plasma did not facilitate the release reaction. Heparin, in contrast to the other anticoagulants used in this study, does not complex Ca^{2+} but may allow access of this cation to the membrane. It was shown that the presence of Ca^{2+} was fundamental to the release process by the following experiments. Platelets were prepared from plasma containing EDTA. Various Ca^{2+} concentrations could then be achieved by adding calcium chloride, the plasma heated at 60 °C and the LD activity determined as described earlier. The results indicate that the release of activity is dependent on the Ca^{2+} concentration, with an increase in the rate of release as the Ca^{2+} concentration increases until an inhibition occurs. This inhibition was due to increased Ca^{2+} concentrations rendering the enzyme more susceptible to heat inactivation and not an effect on the release phenomena. In addition, although the actual concentration of free Ca^{2+} required to mediate the release process is difficult to determine because of the presence of EDTA, it would appear to be 1–6 mM. That calcium has an effect on the release reaction of platelets has been described previously, but it should be pointed out that calcium has also been shown in erythrocytes to affect many membrane functions, including the rate of haemolysis, cation transport, cell size and shape, viscosity and deformability [2]. It therefore seems probable that calcium is involved in the release process at the membrane level. Thus, we believe that the system described has a number of unique properties. It is easily manipulated and investigations of this type may enable a better understanding of the release of enzymes from other types of cells.

References

1 BAIS, R. and EDWARDS, J. B.: Plasma lactate dehydrogenase activity will be increased if detergent and platelets are present. Clin. Chem. *23:* 1056–1058 (1977).

2 DUNN, M. J.: Red blood cell calcium and magnesium: effect upon sodium and potassium transport and cellular morphology. Biochim. biophys. Acta *252:* 97–116 (1974).

3 HOLMSEN, H.: Collagen-induced release of adenosine diphosphate from blood platelets incubated with radioactive phosphate *in vitro*. Scand. J. clin. Lab. Invest. *17:* 239–246 (1965).

4 HOLSEN, H.; DAY, H.J., and STORKMORKEN, H.: The blood platelet release reaction. Scand. J. Haemat. *8:* suppl., p. 1 (1969).

5 ROTHWELL, D.J.; JENDREZEJCZAK, B.; BECKER, M., and DOUMAS, B.I.: Lactate dehydrogenase activities in serum and plasma. Clin. Chem. *22:* 1024–1026 (1976).

6 VALDORF-HANSEN, J.F. and ZUCKER, M.B.: Effect of temperature and inhibitors on serotonin-^{14}C release from human platelets. Am. J. Physiol. *220:* 105–111 (1971).

7 WACKER, W.E.C.; ULMER, D.D., and VALLEE, B.L.: Metalloenzymes and myocardial infarction II. New Engl. J. Med. *225:* 449 (1956).

R. BAIS, PhD, Division of Clinical Chemistry, Institute of Medical and Veterinary Science, *Adelaide, South Australia 5000* (Australia)

Enzymes in Health and Disease. Inaug. Scient. Meet. Int. Soc. Clin. Enzymol., London 1977, pp. 23–26 (Karger, Basel 1978)

Serum CK and LD and Their Isoenzymes in Swine following Exercise and Heat Stress

COLIN BLOOR, FRANK WHITE, T. M. SANDERS, MICHAEL W. NEWTON, JOHN GRIFFITHS and PAUL WOLF

Department of Pathology, University of California, San Diego, and Veterans Administration Hospital, La Jolla, Calif.

A great interest exists in exercise programs for prevention and rehabilitation of patients with coronary artery disease. The miniswine has a coronary artery distribution similar to that of man, but the dog coronary artery distribution is quite different than that of humans. Since swine respond physiologically to exercise stress in a manner similar to man, we employed the miniswine in our exercise training program [6].

It has been established that CK and LD are released from human skeletal muscle with exercise and trauma [10]. The changes in serum enzyme levels after exercise in trained and untrained subjects has been studied. The greatest increment of enzyme increase was observed in the untrained subjects and the least in trained subjects [7].

Increased enzyme levels in the serum are usually associated with necrosis of cells. However, enzyme levels in serum may also increase with muscular exercise, exposure to cold, altitude, hypoxia, vibration, confinement, and exposure to heat stress [1]. Our goal was to ascertain if exercise training would influence the release of the enzymes CK and LD from various organs when the animals were subjected to various forms of environmental stress, including severe running exercise, heat stress or handling stress.

Materials and Methods

Seven adult miniswine were trained for 10 months by daily running at 80% of their maximal heart rate for 1 h. For endurance exercise training of the miniature swine, we used a small enclosed track which has 4 revolving compartments sufficiently large to contain 2 pigs each.

Exercise was conducted daily for 5 days of each week. Initially, the animals were conditioned at mild exercise levels, until, by the fourth week of training, the animals were running 1 h at 6 km/h. Thus, each animal ran about 35 km/week on the circular track. To characterize the effects of endurance exercise training on each animal, individual resting heart rates and individual responses to a standard treadmill stress test were measured.

After training, these animals were compared to 7 of their untrained litter mates by measuring the serum enzymes CK and LD following exercise and heat stress. The heat stress was produced by heating the miniswine in an environmental container equipped with a space electrical heater with the rectal temperature rising in one group to 104 °F and in another group to 109 °F. Rectal temperatures were monitored in all tests. Total CK and LD were determined on the Beckman TR enzyme instrument while the CK and LD isoenzymes were performed utilizing the Helena cellulose acetate electrophoretic procedure [9].

Results

In the untrained miniswine serum CK increased significantly ($p < 0.001$) from 280 ± 40 U/l to 3,150 ± 340 U/l during exercise, a comparable increase occurring with the trained animals where CK rose from 185 ± 51 U/l to 2,595 ± 450 U/l. In both groups rectal temperature was 109 °F during exercise. When animals were heat stressed to 104 and 109 °F without exercise, there was a significant difference ($p < 0.05$) between untrained (3,490 ± 970 U/l) and trained (2,270 ± 440 U/l) miniswine. Total serum LD levels followed a pattern similar to CK but were much less sensitive to exercise or heat stress. The MM isoenzyme of CK showed the major increase after exercise or heat. However, the BB isoenzyme of CK was found in 82% of samples following heavy exercise or high heat and accounted for 13% of the total CK. As for the LD isoenzymes, LD_5 showed the greatest increase.

The normal pattern of LD isoenzymes in miniswine is that LD_1 is greater than LD_2 resembling acute myocardial infarction in humans. Plasma hemoglobin was normal ruling out hemolysis as the cause for the increase in LD_1.

Discussion

This study has demonstrated that in the animal model, the miniswine, CK and LD enzyme elevation occurs with exercise stress as is seen in humans and that exercise generated heat greater than 3 °F accounts for part of the enzyme release. The enzyme elevation was greater in the untrained animal than in the trained animal. Heat stress without exercise also caused a prominent increase in CK and LD in the untrained over that of the trained. A surprising finding was the frequent increase in CK isoenzyme BB. This isoenzyme is usually associated with the central and peripheral nervous system [2]. The animals exhibited no signs of nervous system abnormalities. Thus, the BB isoenzyme may have been released from other organs such as lung, smooth muscle of the gastrointestinal tract or urogenital tract. It has been shown in exercising swine that visceral flow is depressed [5].

The presence of CK and, to a lesser extent, of LD in the serum after exercise occurs in both humans and animals [8]. When the miniswine were confined to a small cage for several hours, serum CK and LD elevations occurred. Thus, serum CK and LD elevation following exercise stress is partially caused by release of the enzymes because of mechanical stress but is also caused by the effect of the heat stress injuring cellular organelles releasing CK and LD.

Marathon running in conditioned subjects causes a rise in total LD, with a rise in LD isoenzymes 3, 4, and 5, but not in LD 1 and 2. Thus, there is no myocardial enzyme release in exercise trained humans [4].

Decrease in serum LD activity immediately after exercise in conditioned humans may be due to more rapid clearance of LD from the serum as a result of increased cardiac output and organ blood flow accompanying endurance exercise [12].

The increase in CK and LD enzyme levels in the swine after exercise or after heat stress suggests altered cellular membrane permeability. The changes in the cell due to anoxia, pH decline and changes in fluid composition lead to alteration in the cell membrane resulting in enzyme release [3].

Thus, CK is a more sensitive enzyme than LD to evaluate acute exercise stress or heat prostration in conditioned individuals. The level of physical activity and temperature should be taken into account to avoid an incorrect diagnosis when one utilizes CK in human patients to assess the possibility of acute myocardial infarction, CNS damage or rhabdomyolysis [11]. This investigation has established that the pig can be exercise-trained and is an appropriate model for endurance studies.

Summary

Seven adult miniswine were trained for 10 months by daily running at 80% of their maximal heart rate for 1 h. After training, these animals were compared to seven of their untrained litter mates by measuring the serum enzymes CK and LD and their isoenzymes following exercise and heat stress. The untrained animals showed a significant increase in CK and LD following exercise or heat stress. The MM isoenzyme of CK showed the major increase after exercise or heat. However, the BB isoenzyme of CK was found in 82% of samples following heavy exercise or when the animals were stressed with high temperature (109 °F without exercise).

References

1 BEDRAK, E.: Blood serum enzyme activity of dogs exposed to heat stress and muscular exercise. J. appl. Physiol. *20:* 587–590 (1965).
2 ITANO, M.: The detection of CPK_1 (BB) in serum: a summary of sixteen cases. Am. J. clin. Path. *65:* 351–355 (1976).
3 REDDY, M.V.V.; KASTENSCHMIDT, L.L.; CASSENS, R.B., and BRISKEY, E.J.: Studies on stress-susceptibility: the relationship between serum enzyme changes and the degree of stress-susceptibility. Life Sci. *10:* 1381–1391 (1971).
4 ROSE, L.E.; BOUSSER, J.E., and COOPER, K.H.: Serum enzymes after marathon running. J. appl. Physiol. *29:* 355–357 (1970).
5 SANDERS, M.; RASMUSSEN, S.; COOPER, D., and BLOOR, C.: Renal and intrarenal blood flow distribution in swine during severe exercise. J. appl. Physiol. *40:* 32–93 (1976).
6 SANDERS, T.M. and BLOOR, C.M.: Effects of endurance exercise on serum enzyme activities in the dog, pig and man. Proc. Soc. exp. Biol. Med. *148:* 823–828 (1975).
7 SANDERS, T.M. and BLOOR, C.M.: Effects of repeated endurance exercise on serum enzyme activities in well-conditioned males. Med. Sci. Sports *7:* 44–47 (1975).
8 WOLF, P.L. and WILLIAMS, D.: Practical clinical enzymology: techniques and interpretations, p. 170 (Wiley & Sons, New York 1973).
9 WOLF, P.L.: Evaluation and utilization of a kinetic enzyme direct measuring photometer. J. clin. Path. *28:* 587–591 (1975).
10 WOLF, P.L.; KEARNS, T.; NEUHOFF, J., and LAURIDSON, J.: Identification of CPK isoenzyme MB in myocardial infarction. Lab. Med. *5:* 48–50 (1974).
11 WOLF, P.L.: Serum enzymes in the diagnosis of acute myocardial infarction. Pract. Cardiol. *1:* 19–21 (1975).
12 WOLFSON, S.; ROSE, L.I.; BOUSSER, J.E.; PARISI, A.F.; ACOSTA, A.E.; COOPER, K.H., and SCHECHTER, E.: Serum enzyme levels during exercise in patients with coronary heart disease: effects of training. Am. Heart J. *84:* 478–483 (1972).

P.L. WOLF, MD, University Medical Center, 225 West Dickinson Street, *San Diego, CA 92103* (USA)

Enzymes in Health and Disease. Inaug. Scient. Meet. Int. Soc. Clin. Enzymol., London 1977, pp. 27–31 (Karger, Basel 1978)

A Slow Alkaline Phosphatase Isoenzyme

P. M. Crofton and A. F. Smith

Department of Clinical Chemistry, Royal Infirmary, Edinburgh

Introduction

Several workers have reported the existence of an atypical band of alkaline phosphatase which moves slowly on polyacrylamide gel electrophoresis [2]. We have studied the properties of this slow band in the serum of 8 patients with the aim of determining its nature and clinical significance.

Patients and Methods

From a total of over 2,000 alkaline phosphatase isoenzyme analyses, sera from 8 patients possessed the slow band as their major component on polyacrylamide gel electrophoresis. Several other sera had bands with similar mobility but these bands were always fainter. Most of the patients studied had diseases mainly affecting the intestine or the lung, but this was not always so.

Alkaline phosphatase activity was measured using as substrate 14 mmol/l *p*-nitrophenyl phosphate in 0.9 mol/l diethanolamine buffer pH 10.2 containing 0.5 mmol/l magnesium chloride. Rates of reaction were continuously monitored at 410 nm at 37 °C on an LKB 8600 reaction rate analyser. Inhibition studies were carried out in the presence of 5 mmol/l inhibitor. Heat inactivation curves were determined on whole sera by the method of Whitby and Moss [4].

Polyacrylamide gel electrophoresis of alkaline phosphatase was carried out using a modification of the method of Kaplan and Rogers [1]. In the inhibitor studies the gel was stained in the presence of 5 mmol/l inhibitor and compared with a control gel stained in the absence of inhibitor.

Column chromatography was carried out on a pre-calibrated 2.5 × 100 cm Sepharose 6B column equilibrated with 0.1 mol/l Tris-HCl pH 7.7 containing 0.05 mol/l NaCl.

Immunoelectrophoresis was performed as described by Scheidegger [3]. The precipitin arcs were stained for alkaline phosphatase activity according to the method of Kaplan and Rogers [1].

Results

The qualitative effects of inhibition by L-homoarginine and L-phenylalanine were studied. The slow band in all 8 patients behaved qualitatively like the liver and bone isoenzymes rather than the intestinal isoenzyme in this respect. Quantitative studies on whole sera supported these findings.

The fall in activity which resulted on heating a patient serum at 56 °C was measured. It appears that in this respect the slow band exhibits behaviour intermediate between that of the liver and bone isoenzymes.

Sepharose 6B chromatography suggested molecular weights of 550,000 and 220,000 for the slow band and liver alkaline phosphatase, respectively.

On immunoelectrophoresis, alkaline phosphatase activity was demonstrable in the IgG precipitin arc in all 8 patient sera. No alkaline phosphatase activity was present in the IgA precipitin arcs of these sera, nor was there any enzyme activity in any of the precipitin arcs formed by a control serum containing only the liver isoenzyme.

Supernatants from the following incubations of sera with antisera were

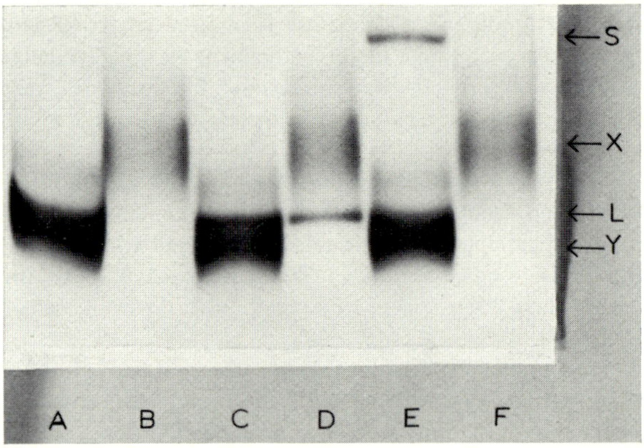

Fig. 1. Alkaline phosphatase electrophoresis patterns on polyacrylamide gel. A = IgA antiserum control; B = IgG antiserum control; C = supernatant from incubation a; D = supernatant from incubation b; E = supernatant from incubation c; F = supernatant from incubation d; S = slow band; X = IgG antiserum alkaline phosphatase; L = liver isoenzyme; Y = IgA antiserum alkaline phosphatase. See text for details.

subjected to polyacrylamide gel electrophoresis (fig. 1): (a) liver isoenzyme serum (control) + IgA antiserum; (b) liver isoenzyme serum (control) + IgG antiserum; (c) slow band serum + IgA antiserum, and (d) slow band serum + IgG antiserum.

The results showed that the slow band had disappeared from the supernatant of incubation d. Immunoelectrophoresis showed that the precipitin arc against IgG antiserum had also disappeared. The slow band was unaffected by anti-IgA (c). The liver isoenzyme was unaffected by either anti-IgG (b) or anti-IgA (a).

A slow band serum was incubated with equal volumes of other sera containing (a) liver, (b) bone and (c) intestinal alkaline phosphatase, respectively. All sera were also incubated with the appropriate volume of saline as controls. Subsequent electrophoresis on polyacrylamide gel (fig. 2) showed that in incubations a and b the slow band had been augmented and the liver and bone bands respectively had been diminished compared with the controls. However, in incubation c, no change was observed in the intestinal band compared with the control.

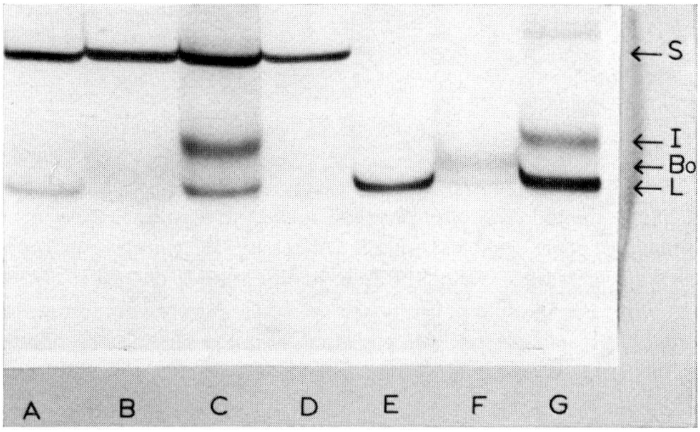

Fig. 2. Alkaline phosphatase electrophoresis patterns on polyacrylamide gel. A = Slow band serum + liver isoenzyme serum; B = slow band serum + bone isoenzyme serum; C = slow band serum + intestinal isoenzyme serum; D = slow band serum control; E = liver isoenzyme serum control; F = bone isoenzyme serum control; G = intestinal isoenzyme serum control; S = slow band; I = intestinal band; Bo = bone band; L = liver band. See text for details.

Discussion

The slow band complexes in the 8 sera studied were similar in most of their properties. The inhibition studies showed that the complex resembled liver or bone alkaline phosphatase rather than the intestinal isoenzyme. The intermediate heat stability shown by the slow band complex could be due to either increased stability of the bone isoenzyme or reduced stability of the liver isoenzyme.

All 8 sera studied possessed an alkaline phosphatase-positive precipitin arc against IgG antiserum on immunoelectrophoresis. In different sera, this was of variable light chain type. Removal of IgG from a patient's serum by immunoprecipitation was accompanied by a concomitant removal of the slow band. This suggests that the slow band consists of an alkaline phosphatase-IgG complex.

The molecular weight of the slow band complex (550,000) suggests that it may consist of two IgG molecules (molecular weight 160,000) associated with one liver alkaline phosphatase molecule (molecular weight 220,000).

It appears that a substance in sera containing the slow band can bind liver and bone alkaline phosphatase from other sera. No evidence was found of a corresponding ability to bind intestinal alkaline phosphatase. It may be postulated that this substance is an unusual IgG molecule and that the binding occurs as an antibody-antigen type reaction *in vivo*.

Summary

Sera from 8 patients with a marked slow-moving alkaline phosphatase band on electrophoresis were investigated. Inhibitor studies showed that in each case this band resembled liver or bone, but never intestinal, alkaline phosphatase. Immunoelectrophoretic and molecular weight studies indicate that the slow band complex consisted of two IgG molecules associated with one alkaline phosphatase molecule. Serum containing the slow band was able to bind liver or bone, but not intestinal, alkaline phosphatase from other sera to form the slow band complex. No clinical condition was found common to all 8 patients.

References

1 KAPLAN, M.M. and ROGERS, L.: Separation of human serum alkaline phosphatase isoenzymes by polyacrylamide gel electrophoresis. Lancet *ii:* 1029–1031 (1969).

2 Qirbi, A.A. and Moss, D.W.: Incidence and some properties of an electrophoretically slow form of alkaline phosphatase in sera of patients with disease of the intestine. Clin. chim. Acta *60:* 1–6 (1975).

3 Scheidegger, J.J.: Une micro-méthode de l'immune-électrophorèse. Int. Archs Allergy appl. Immun. *7:* 103–110 (1955).

4 Whitby, L.G. and Moss, D.W.: Analysis of heat inactivation curves of alkaline phosphatase isoenzymes in serum. Clin. chim. Acta *59:* 361–367 (1975).

A.F. Smith, MD, Department of Clinical Chemistry, Royal Infirmary, *Edinburgh EH3 9YW* (Scotland)

Enzymes in Health and Disease. Inaug. Scient. Meet. Int. Soc. Clin. Enzymol., London 1977, pp. 32–37 (Karger, Basel 1978)

The Fate of Liver Alkaline Phosphatase in the Body[1]

S. MASSARRAT, V. HERBERT and D. SCHAEFER

Medical Policlinic and Department of Surgery of Philipps-University, Marburg

The very intensive experimental investigations on alkaline phosphatase (AP) revealed without a doubt that serum AP originates mainly from the liver, especially in bile duct disorders [3, 5–8]. However, the catabolic sites and the elimination of the enzyme through the bile duct remained the subject of controversy [1, 2, 4, 6, 11]. In an attempt to answer these questions, we isolated the enzyme from the livers of 5 dogs, whose main bile ducts were ligated for 1 day. The purification was performed with some modifications according to SUSSMAN *et al.* [9]. The enzyme could be purified up to 5,440 times. In polyacrylamide gel electrophoresis, the enzyme was revealed as a single band, visualized by protein staining and enzyme activity staining with α-naphthyl-phosphate. With rabbit antibodies, obtained by immunization with pure enzyme, the enzyme band, first eletrophoresed on polyacrylamide gel, then diffused in the surrounding agar gel against antibodies, shows only one precipitation line, detected by enzyme activity with α-naphthyl-phosphate. The enzyme was labelled with ^{125}-iodine with the lactoperoxidase method, described by THORELL and JOHANSSON [10].

The specific activity of labelled protein amounted to 130–300 μCi/mg protein. Neither loss of enzyme activity, nor change of electrophoretic mobility or immune precipitability could be observed after this labelling procedure.

In 4 dogs, the bile flow was completely deviated by an extracorporeal drainage system, in which the bile volume could be measured continuously without interruption of the enterohepatic circulation. 30–100 μg of labelled

[1] Supported by Deutsche Forschungsgemeinschaft, Bad Godesberg (FRG).

AP were injected intravenously into 10 dogs each weighing about 10 kg, 4 of which had extracorporeal deviation of bile flow. Two dogs were investigated before as well as 1 week after complete ligation of the main bile duct. Blood, and bile samples were taken intermittently over 2–4 days. In the first 12–16 h after injection, the bile volumes were measured continuously. After taking bile samples, the remainder was reinfused to the distal part of the bile duct. In 4 dogs, the radioactivity was counted intermittently by colimator placed over heart, liver and kidney. Total bound radioactivity, protein-bound radioactivity and the radioactivity related to washed immune precipitates of sera and bile samples after 2 days' incubation with antibodies were measured.

Figure 1 shows the course of PBI radioactivity in the sera and bile of one animal. The radioactivity is removed from the sera abruptly in the first 15 min after injection. This was followed in turn by an increase within ½ to 1 h with a logarithmic fall of PBI afterwards. At the same time, where the PBI increases in the sera the radioactivity appears in the bile and reaches its maximum and falls more abrupt than in the sera to a very low level.

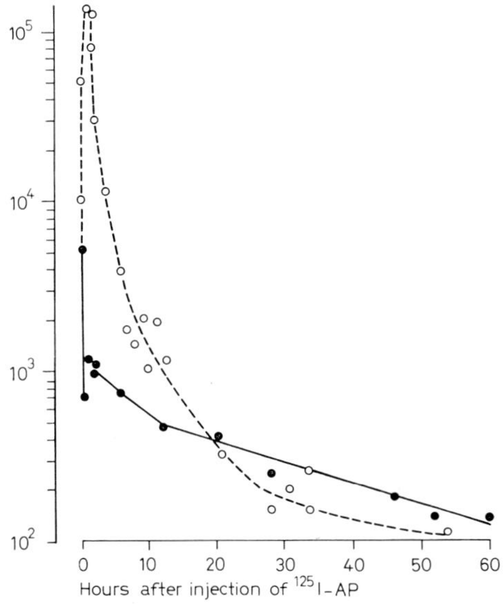

Fig. 1. The disappearance rate of protein-binding radioactivity in serum (●) and bile (○) of a dog after intravenous injection of ^{125}I-AP.

Figure 2 shows the mean values of PBI in the sera of 8 dogs and the bile samples of 4 dogs during the first 8 h (the upper portion of fig. 2). As the PBI radioactivity decreases in the sera soon after injection, the percentage of PBI radioactivity, related to total radioactivity, decreases from 87% during the first minutes to 43% after 15 min. The percentage of PBI in the bile amounts to 60–70% in the first 30–90 min and falls to less than 40% after 6 h as is the case with sera.

While PBI radioactivity decreases in the sera soon after intravenous injection, radioactivity counted by colimator appears selectively in the liver. During the first minutes, it is 20 times lower in the heart and the kidney in comparison to the liver. While radioactivity falls in the liver, it increases in the heart and the kidney during 2 h after injection and remained afterwards almost at the same level over these organs. The percentage of PBI radioactivity, which is precipitated with antibodies, in sera and bile specimens taken from 3 animals is shown in figure 3. The immune precipitability

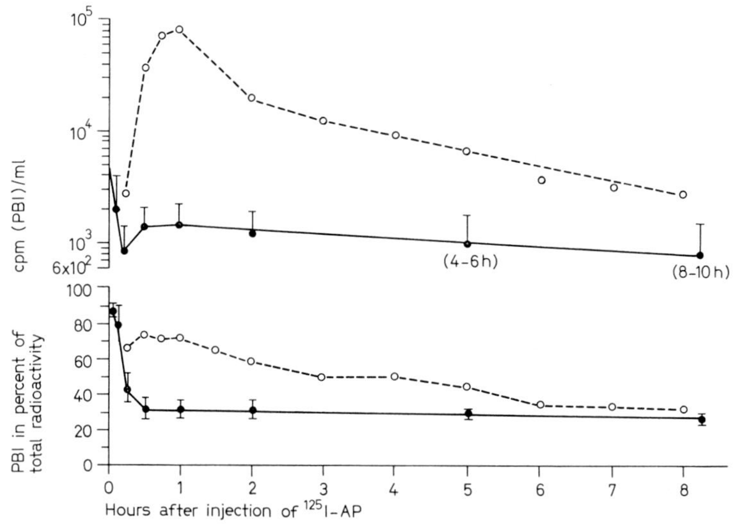

Fig. 2. Upper graph: The disappearance of protein-binding radioactivity in sera (●, n=8) and bile specimens (○, n=4) of the dogs. Lower graph: The course of the relation between protein-binding radioactivity and total radioactivity in the sera (●) and bile specimens (○) of the dogs after intravenous injection of ^{125}I-AP.

decreases rapidly from 81% within the first minutes to 25% after 30 min and decreases later to less than 20% in the sera. About 25–35% of PBI in the bile is precipitable with antibody, which falls later to less than 20% as is the case with sera. The columns at the right of figure 3 show the immune precipitability of AP. The precipitability of labelled AP, as measured by radioactivity, is 95%; that in sera with obstructive jaundice, as measured by enzyme activity, is 96%; in canine sera without jaundice 56%; in bile 91%, and in butanol extract of liver 76%.

Table I shows the total bile flow and excretion rate of PBI through the main bile duct during the first 15 h separated in three 5-hour periods, after intravenous injection of labelled AP. About 4.19% of injected PBI was eliminated in 55.4 ml bile, as an average during this time.

Neither in the course of the abrupt decrease and the following increase nor in the later course of the logarithmic fall of PBI could any difference be observed between dogs with and without obstructive jaundice; the half-life

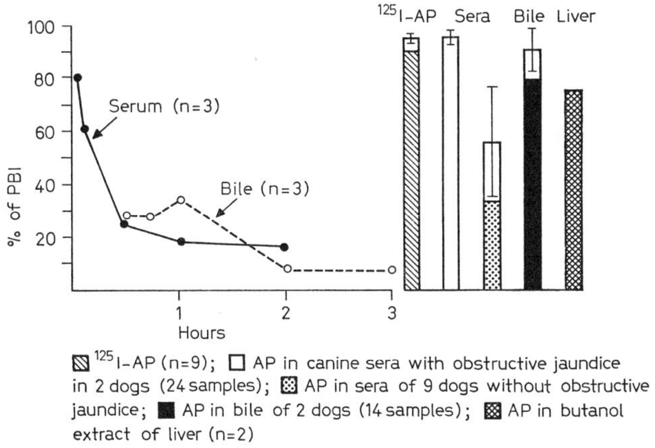

Fig. 3. Left: Percentage of precipitability of protein-binding radioactivity in the sera and bile specimens of 3 dogs with specific rabbit antibody after intravenous injection of ^{125}I-AP. Right: From left to right the columns represent the percentage of precipitability of labelled AP (by measuring the radioactivity in the immune precipitates) and enzyme activity of AP in the sera of dogs with obstructive jaundice (white column), in the sera of the dogs without obstructive jaundice, and in bile specimens and butanol extract of liver (by measuring the enzyme activity in the supernatants after precipitation with specific antibody).

Table I. Total bile flow and excretion rate of PBI through the main bile duct during the first 15 h

	Hours			
	0–5	6–10	11–15	total
Bile flow, ml	21.6	19.1	14.7	55.4
Excretion rate, %[1]	3.54	0.37	0.18	4.09

[1] The values are expressed in relation to the injected ^{125}I-AP in 4 dogs (averages).

of PBI^{125} amounts in one dog to 10.5 h before and 11 h after ligation of the common bile duct, in the second dog to 28 and 20 h, respectively.

Our results suggest that AP is stored selectively in the liver soon after intravenous injection and is mainly catabolized there. The excretion rate of the enzyme through the bile duct is very low and can be neglected. The question arises, whether the purified enzyme is identical to the enzyme which appears in the sera. As the antibody against the pure enzyme precipitates almost completely the enzyme activity in the sera of obstructive jaundice and the bile, it is improbable that the molecular structure of the purified enzyme differs from that of the enzyme observed in the sera.

References

1 CANTAROW, A. and MILLER, L. L.: Non-excretion of jaundice serum alkaline phosphatase in bile of normal dogs. Am. J. Physiol. *153:* 444–446 (1948).

2 CLUBB, J. S.; NEALE, F. C., and POSEN, S.: The behavior of infused human placental alkaline phosphatase in human subjects. J. Lab. clin. Med. *66:* 493–507 (1965).

3 KAPLAN, M. M. and RIGHETTI, A.: Induction of rat liver alkaline phosphatase: the mechanism of the serum elevation in bile duct obstruction. J. clin. Invest. *49:* 508–516 (1970).

4 VEEN, H. H. LE; TALBOT, L. J.; RESTUCCIA, M., and BARBERIO, J. R.: Metabolism and excretion of alkaline phosphatase: relation to liver function and determination of maximal secretory rates of liver. J. Lab. clin. Med. *36:* 192–205 (1950).

5 MASSARRAT, S.: Alkaline phosphatase in the subcellular fractions of normal and damaged liver. Enzyme *12:* 402–408 (1971).

6 MASSARRAT, S. and MASSARRAT, Sch.: Origin of serum alkaline phosphatase in hepatobiliary diseases. A metabolic study. Enzyme *12:* 593–605 (1971).

7 POLIN, S.G.; SPELLBERG, M.A.; TEITELMAN, L., and OKUMURA, M.: The origin of elevation of serum alkaline phosphatase in hepatic disease. Gastroenterology *42:* 431 (1962).

8 SEBESTA, D.G.; BRADSHAW, F.J., and PROCKOP, D.J.: Source of the elevated serum alkaline phosphatase activity in biliary obstruction: studies utilizing isolated liver perfusion. Gastroenterology *47:* 166–170 (1964).

9 SUSSMAN, H.H.; SMALL, P.A., and COTLOVE, E.: Human alkaline phosphatase: immunochemical identification of organ-specific isoenzymes. J. biol. chem. *243:* 160–166 (1968).

10 THORELL, J.I. and JOHANSSON, B.G.: Enzymatic iodination of polypeptides with ^{125}I to high specific activity. Biochim. biophys. Acta *251:* 363–369 (1971).

11 WANG, C.C. and GROSSMAN, M.I.: Non excretion of serum alkaline phosphatase in the liver and pancreas of normal dogs. Am. J. Physiol. *156:* 256–260 (1949).

S. MASSARRAT, MD, Medical Policlinic, Philipps-University, *D-3550 Marburg/Lahn* (FRG)

Enzymes in Health and Disease. Inaug. Scient. Meet. Int. Soc. Clin. Enzymol., London 1977, pp. 38–45 (Karger, Basel 1978)

Human Tissue Analysis by Analytical Subcellular Fractionation Techniques in Combination with Enzymic Microassays[1]

T. J. PETERS and CAROL A. SEYMOUR

Department of Medicine, Royal Postgraduate Medical School, London

Clinical enzymologists have hitherto concentrated on the assay of enzyme levels in the circulating plasma and they have been able to derive a considerable amount of information about the pathological processes in certain organs, e.g. heart, liver. There are however important organs, e.g. intestine, brain, where disease processes only occasionally produce abnormalities in the plasma enzymes. A logical development is therefore to study the levels of enzymes in the diseased tissue itself. The assay of tissue enzymes is a well recognised technique for demonstrating specific deficiencies in various syndromes, e.g. lysosomal storage diseases, but little information is available on enzyme levels in other diseases, particularly those chronic diseases where the pathogenesis is ill-understood. We have therefore begun to apply analytical techniques [PETERS, 1977] to closed biopsy specimens from a variety of diseased organs.

As it was planned to assay a wide spectrum of enzymes, highly sensitive assays employing fluorigenic and radiolabelled substrates were used [PETERS, 1976; SEYMOUR and PETERS, 1977]. Table I shows the levels of various enzymes in 10 needle biopsy specimens from a single monkey liver. There is a 2- to 3-fold variation in the activity: duplicate assays of the same specimen agreed to within 5%. It is likely however that in diseased tissue, e.g. cirrhosis, the variation would be greater.

An important question which was not considered in previous studies [SCHMIDT and SCHMIDT, 1970] is the subcellular localisation of the various

[1] Supported by the Medical Research Council and The Wellcome Trust.

Table I. Enzyme activities in multiple biopsies from monkey liver

Enzyme	Specific activity, mU/mg protein (n = 10)	
	mean ± SE	range
Plasma membrane		
5′-Nucleotidase	12.3 ± 2.1	8.31–16.3
Alkaline phosphatase	1.32 ± 0.48	0.91–2.52
Leucyl-2-naphthylamidase	2.10 ± 0.33	1.61–2.72
Biliary canaliculi		
γ-Glutamyl transpeptidase	6.27 ± 0.51	5.37–6.78
Endoplasmic reticulum		
Neutral α-glucosidase	0.91 ± 0.20	0.43–1.10
Lysosomes		
N-Acetyl-β-glucosaminidase	3.43 ± 0.34	2.71–3.77
Acid phosphatase	9.11 ± 0.75	7.67–10.1

enzymes. Assay of levels in whole homogenates may obscure a selective quantitative or qualitative alteration in an enzyme in a particular compartment.

Liver Disease

In a series of patients with chronic liver disease, mainly due to alcoholism, primary biliary cirrhosis and chronic active hepatitis, assays of marker enzymes for the principal subcellular organelles were performed concurrently on plasma samples and liver biopsy homogenates. Table II shows the correlation between these activities. For 2 of the 3 plasma membrane marker enzymes there is a significant *positive* correlation between the tissue and plasma levels. Leucyl-2-naphthylamidase, although having a significant plasma membrane localisation, has cytosol, lysosomal and endoplasmic reticulum components. It is likely that if the plasma membrane component alone were assayed, a correlation with the plasma levels would be achieved for this enzyme. Enzymes associated with other organelles do not show this correlation. It is thus clear that an increased plasma level of alkaline phosphatase and 5′-nucleotidase reflects an increased liver concentration of these enzymes.

Figure 1 shows the result of an analytical subcellular fractionation

Table II. Correlation between hepatic and serum enzyme levels in chronic liver disease

Enzyme	Correlation coefficient	
Plasma membrane		
5′-Nucleotidase	+0.70	$p<0.001$
Alkaline phosphatase	+0.56	$p<0.01$
Leucyl-2-naphthylamidase	−0.12	n.s.
Biliary canaliculi		
γ-Glutamyl transpeptidase	+0.32	n.s.
Endoplasmic reticulum		
Neutral α-glucosidase	+0.18	n.s.
Lysosomes		
N-Acetyl-β-glucosaminidase	+0.30	n.s.
Acid phosphatase	+0.07	n.s.

study of a fragment of liver from a control subject. There is clear resolution of the various organelles: the mitochondria and peroxisomes (reflected by the malate dehydrogenase and catalase, respectively) have similar equilibrium densities to that found in rat liver [PETERS and SHIO, 1976]. However, *N*-acetyl-β-glucosaminidase, the lysosomal marker, shows a major soluble component and the lysosomes show a broader density distribution with a lower modal density (1.17 compared to 1.23) than rat liver. This study also demonstrates differences in the distribution of the putative plasma membrane markers. 5′-Nucleotidase has a distinct peak at an equilibrium density of 1.12, alkaline phosphatase has a similar distribution with a broader peak and more activity deeper in the gradient. Leucyl-2-naphthylamidase (not shown) has more soluble activity remaining in the sample layer and considerable activity deep in the gradient [SEYMOUR *et al.*, 1974]. This heterogeneity agrees well with the different degree of correlation between tissue and plasma activities.

These studies indicate how subcellular fractionation procedures used in conjunction with the enzyme assays can yield valuable information about hepatic disease. Present studies are using these techniques to investigate organelle pathology in iron overload [PETERS and SEYMOUR, 1976; PETERS *et al.*, 1977a], alcoholic liver disease [SEYMOUR *et al.*, 1975; JENKINS and PETERS, 1978], liver tumours [SEGAL *et al.*, 1976; PETERS *et al.*, 1977b] and the Dubin-Johnson syndrome [SEYMOUR *et al.*, 1977; PETERS and SEYMOUR, 1978].

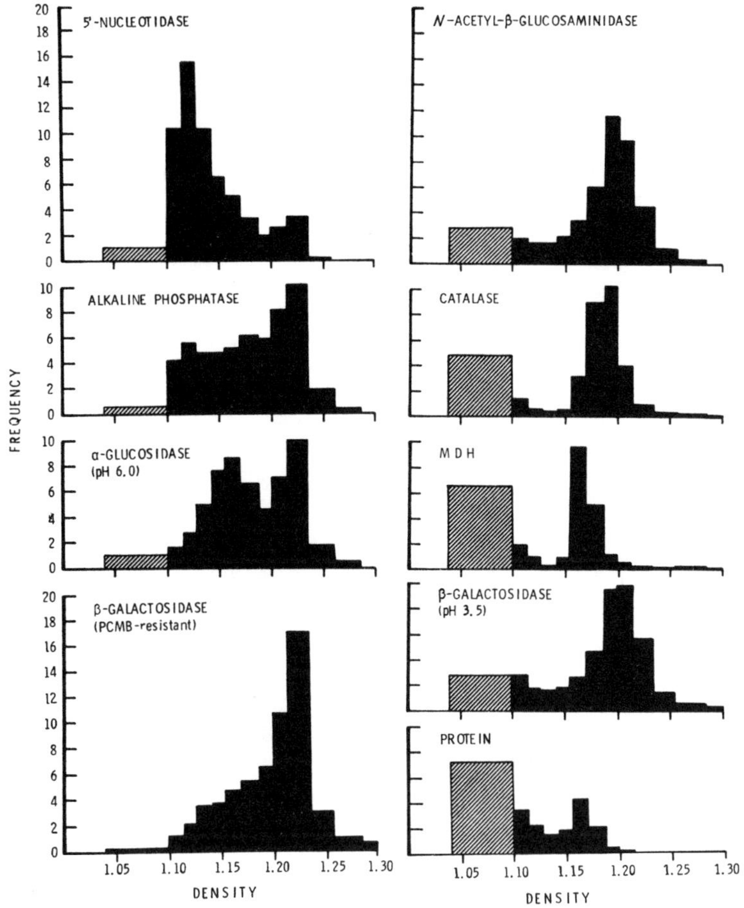

Fig. 1. Isopycnic centrifugation of post nuclear supernatant from liver biopsy homogenate from 24-year-old male showing no histological abnormality. Figure shows frequency-density distribution for marker enzymes and protein. Frequency is defined as fraction of total recovered enzyme activity present in gradient fraction divided by density span covered. The hatched area indicates, over an arbitrary abscissa interval, activity remaining in the sample layer and presumed to represent the soluble components. Reproduced from SEYMOUR *et al.* [1974].

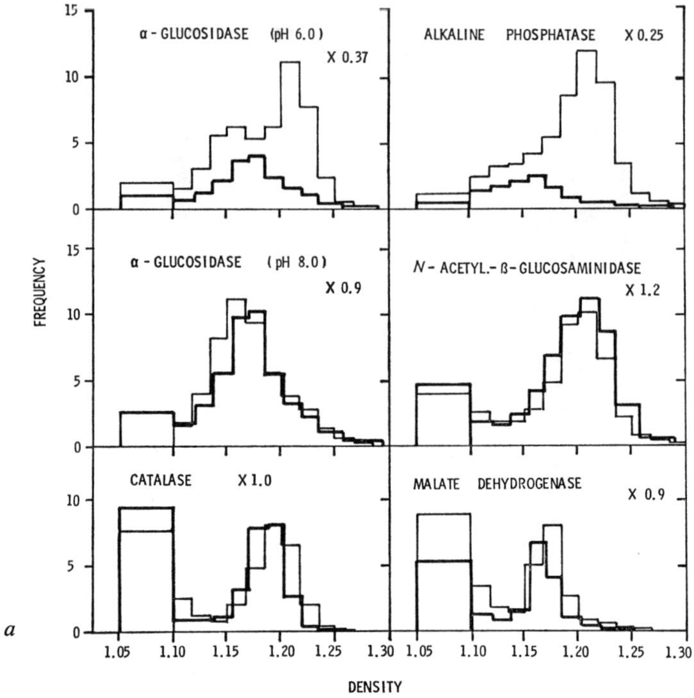

Fig. 2. a Isopycnic centrifugation of post nuclear supernatant from jejunal biopsy homogenate from 33-year-old female (▬) with untreated coeliac disease compared with control subjects (—) [PETERS, 1976]. The biopsy showed subtotal villous atrophy. Graph shows relative frequency-density histograms for marker enzymes for the various subcellular organelles: α-glucosidase (pH 6.0) and alkaline phosphatase (brush border); α-glucosidase (pH 8.0) (endoplasmic reticulum); *N*-acetyl-β-glucosaminidase (lysosomes); catalase (peroxisomes); malate dehydrogenase (mitochondria). The distributions are adjusted so that area under the histograms reflects the relative specific activity of the patient's homogenate compared with controls. Other details as figure 1. *b* Isopycnic centrifugation of biopsy extract from same patient after 15 months on a strict gluten-free diet. The biopsy was histologically normal. Other details as figure 2a.

Coeliac Disease

Clinical enzymologists, assaying plasma enzymes, contribute little to our understanding of the fundamental disease process although the consequences of the malabsorption may produce plasma enzymic changes.

Although the light microscopic abnormalities and the response to removal of gluten from the diet have been well described in coeliac disease,

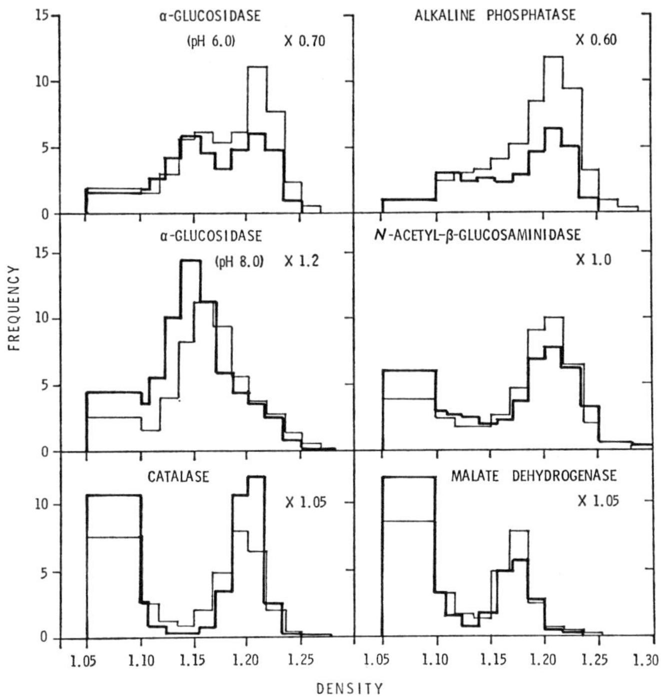

b

little is known of the pathogenesis of the lesion. Ultrastructural studies of the jejunal mucosa from patients with coeliac disease have described a bewildering variety of abnormalities in the subcellular organelles [SHINER, 1967]. In order to evaluate quantitatively these changes the techniques of analytical subcellular fractionation and enzymic microassay have been combined in a study of coeliac disease [PETERS *et al.*, 1978]. In figure 2a, from a patient with untreated coeliac disease, it is clear that the total levels of alkaline phosphatase and α-glucosidase are reduced in the patient. The residual enzymic activity has a very abnormal distribution in the gradient with the complete loss of the brush border peak at density 1.21. Apart from an increase in the total level of the lysosomal enzyme *N*-acetyl-β-glucosaminidase, the other subcellular organelles are relatively unaffected by the coeliac process. In figure 2b, the results of a second fractionation, performed after the patient had taken a strict gluten-free diet for 15 months, are shown. Histologically, the biopsy was indistinguishable from normal. The levels of alkaline phosphatase and α-glucosidase have returned towards normal levels. The fractionation studies show a restoration of the brush border peak at density 1.21 but the level of this component is significantly below normal. This result indicates a persisting brush border abnormality in patients with apparently full recovery from gluten-sensitive coeliac disease. Whether this reflects an underlying defect in the coeliac mucosa remains to be determined [PETERS *et al.*, 1978].

These 2 examples show how tissue enzymic analysis, especially when combined with a subcellular fractionation technique, can lead to a deeper understanding of disease processes. Clearly, any tissue amenable to biopsy can be studied in this fashion and it is hoped that useful progress will be made in other fields.

References

JENKINS, W.J. and PETERS, T.J.: Mitochondrial enzyme activities in liver biopsies from patients with alcoholic liver disease. Gut (in press, 1978).

PETERS, T.J.: Analytical subcellular fractionation of jejunal biopsy specimens: methodology and characterisation of the organelles in normal tissue. Clin. Sci. mol. Med. *51:* 557–574 (1976).

PETERS, T.J.: Application of analytical subcellular fractionation techniques and tissue enzymic analysis to the study of human pathology. Clin. Sci. mol. Med. *53:* 503–511 (1977).

PETERS, T.J.; JONES, P.E., and WELLS, G.: Analytical subcellular fractionation of jejunal biopsy specimens: enzyme activities, organelle pathology and response to gluten withdrawal in patients with coeliac disease. Clin. Sci. mol. Med. (in press, 1978).

PETERS, T.J.; SELDEN, C., and SEYMOUR, C.A.: Lysosomal disruption in the pathogenesis of hepatic damage in primary and secondary haemochromatosis. Ciba Found. Symp. Iron *51:* 317–329 (1977a).

PETERS, T.J. and SEYMOUR, C.A.: Acid hydrolase activities and lysosomal integrity in liver biopsies from patients with iron overload. Clin. Sci. mol. Med. *50:* 75–78 (1976).

PETERS, T.J. and SEYMOUR, C.A.: The organelle pathology and demonstration of mitochondrial superoxide dismutase deficiency in two patients with Dubin-Johnson-Sprinz syndrome. Clin. Sci. mol. Med. (in press, 1978).

PETERS, T.J.; SEYMOUR, C.A.; WELLS, G.; FAKUNLE, F., and NEALE, G.: γ-Glutamyl transferase levels in ascitic fluid and liver tissue. Br. med. J. *i:* 1576–1577 (1977b).

PETERS, T.J. and SHIO, H.: Subcellular fractionation studies on isolated rat jejunal enterocytes with special reference to the separation of lysosomes, peroxisomes and mitochondria. Clin. Sci. mol. Med. *50:* 355–366 (1976).

SCHMIDT, E. and SCHMIDT, F.W.: Enzyme activities in human liver. Enzymol. biol. clin. *11:* 67–129 (1970).

SEGAL, A.W.; GREGORIADIS, G.; LAVENDER, J.P.: TARIN, D., and PETERS, T.J.: Tissue and hepatic subcellular distribution of liposomes containing bleomycin after intravenous administration to patients with neoplasms. Clin. Sci. mol. Med. *51:* 421–425 (1976).

Seymour, C.A.; Neale, G., and Peters, T.J.: Analytical subcellular fractionation studies on human liver biopsies. Biochem. Soc. Trans. *2:* 1101–1104 (1974).

Seymour, C.A.; Neale, G., and Peters, T.J.: Hepatic enzyme changes in alcoholic liver disease. Gut *16:* 839 (1975).

Seymour, C.A.; Neale, G., and Peters, T.J.: Lysosomal changes in liver tissue from patients with the Dubin-Johnson-Sprinz syndrome. Clin. Sci. mol. Med. *52:* 241–248 (1977).

Seymour, C.A. and Peters, T.J.: Enzyme activities in human liver biopsies: assay methods and activities of some lysosomal and membrane bound enzymes in control tissue and serum. Clin. Sci. mol. Med. *52:* 229–239 (1977).

Shiner, M.: Ultrastructure of jejunal surface epithelium in untreated idiopathic steatorrhoea. Br. med. Bull. *23:* 223–227 (1967).

T.J. Peters, PhD, Department of Medicine, Royal Postgraduate Medical School, *London W12 0H5* (England)

Enzymes in Health and Disease. Inaug. Scient. Meet. Int. Soc. Clin. Enzymol., London 1977, pp. 46–49 (Karger, Basel 1978)

Distribution of γ-Glutamyl Transferase in Some Domestic Animals

J. P. Braun, A. G. Rico, P. Benard and V. Burgat[1]

Ecole Nationale Vétérinaire, Laboratoire de Biochimie Médicale, Toulouse

Introduction

In man, γ-glutamyl transferase (GGT, EC 2.3.2.2) is mainly in kidney, liver and pancreas. Thus, serum GGT is widely used as a test of hepatic and pancreatic diseases [6], and urine GGT seems to be a good test of kidney disease [2].

Some reports, using histochemical procedures [1], have indicated that in some animal species, GGT was also a kidney and liver enzyme. In order to apply this enzymatic test to domestic animals, we studied the distribution of this enzyme in the tissues of some animals.

Materials and Methods

Ten animals of each species were used for this study after killing by exsanguination either at the slaughter-house or at our laboratory: horse, swine, cow, calf, ewe, lamb, 1-week-old kitten, baboon and macaque (for kitten and baboon there were only 4 animals).

In each animal, a sample of the following tissues was collected: kidney (cortex and medulla), liver, pancreas, spleen, lung, heart, muscle, stomach and intestine. It was rinsed in saline solution (9 g NaCl/l), blotted and frozen at –30 °C until analyzed within a week. 1 g of each tissue sample was homogenized in a Potter-Elvejhem device with 4 ml of ice-cold saline.

The homogenate was then centrifuged at 3,000 *g* for 5 min and GGT activity measured in the supernatant by use of a commercial kinetic procedure at 30 °C (Boehringer-

[1] The authors wish to thank M. B. Fantini and R. Saez for valuable technical assistance.

France). Moreover, for kidney and liver of horse, swine, ewe and lamb, GGT activity was also measured in the 10,000- and 120,000-*g* supernatants in order to study cell distribution of the enzyme.

Results

Tissue Distribution

As shown in table I, in each species, GGT is mainly in kidney but its specific activity greatly differs: it ranges from 16 U/g fresh tissue in swine to 81 U/g in macaque monkey. Then, pancreas has also a relatively high GGT activity, more particularly in the pig. For liver, all animals except kitten, which has no liver GGT, exhibit an activity ranging from 4 to 9% of that of kidney. All other organs have low GGT levels and tissue distribution is roughly similar in most species.

Cell Distribution

As shown in table II, it can be observed that cytosoluble GGT (120,000-*g* supernatant) is very limited in both kidney and liver; but this fraction is always smaller in the kidney. Most GGT is included in cell structures, probably mitochondrial and microsomal fractions.

Table I. Distribution of GGT in tissues of some domestic animals, %

	Horse	Cow	Calf	Ewe	Lamb	Swine	Kitten	Baboon	Macaque
Kidney cortex[1]	39	61	44	67	67	16	2	72	81
Kidney cortex	100	100	100	100	100	100	100	100	100
Kidney medulla	–	78.6	74.5	24.3	34.2	–	44.1	56.1	53.0
Liver	6.2	8.2	7.3	8.7	8.9	16.4	0	4.2	4.1
Pancreas	17.6	36.7	45.9	8.0	11.6	52.2	18.3	73.6	47.1
Spleen	0.5	0.5	1.6	1.7	1.0	16.5	–	–	2.2
Lung	0.5	1.5	1.9	5.4	9.8	–	–	–	0.8
Heart	0.3	<0.1	<0.1	0.1	<0.1	0.1	0.8	0.1	0.5
Muscle	0.7	<0.1	<0.1	<0.1	<0.1	0.1	1.1	0.4	0.2
Stomach/rumen	–	0.1	0.3	1.0	0.6	0.4	1.3	0.5	1.2
Intestine	1.5	1.0	1.0	1.1	1.3	8.9	21.4	2.4	1.9

[1] U/g.

Table II. Cell distribution of GGT (results are percent of the 3,000-g supernatant activity)

		Supernatant at		
		3,000 *g*	10,000 *g*	120,000 *g*
Horse	liver	100	24.8±12.2	14.7±2.4
	kidney	100	4.9±2.8	3.2±2.1
Swine	liver	100	45.6±10.8	23.3±7.4
	kidney	100	31.3±10.1	5.0±1.3
Ewe	liver	100	23.8±6.6	3.9±2.4
	kidney	100	11.4±2.3	1.4±0.8
Lamb	liver	100	26.0±4.4	2.8±1.7
	kidney	100	15.2±2.8	1.1±0.1

Discussion – Conclusion

Tissue distribution of GGT in animals is roughly similar to that observed in man [5] except that some species, e.g. kitten, have no detectable liver enzyme. Its cell distribution is also similar; main GGT activity is bound in mitochondria and microsomes.

The relative hepatic specificity of this enzyme indicates that for most animals, serum GGT is a good test of hepatic disease; some clinical reports confirm this hypothesis. The high level of GGT in the kidney indicates that urine GGT could be a very good test of kidney disease or kidney toxic lesions [7]. Moreover, it has been shown that liver enzyme is inducible by such drugs as phenobarbital or ethanol in mouse and rat [3, 4], thus this test could be very useful in toxicological studies.

In veterinary medicine or toxicology, the measure of GGT activity in blood, urine and tissues is probably one of the most valuable tests of liver and kidney diseases and enzyme drug induction.

References

1 ALBERT, Z.: Further studies on the activity of gamma-glutamyl transpeptidase in man and other mammals. Enzyme histochemistry. Simposio Antonio Baselli, Milano 1967 (Succ. Fusi Editrice, Pavia 1968).

2 BECK, P. R.: A study of urinary gamma-glutamyl transpeptidase with special reference to renal disease; MD thesis of science (clinical biochemistry), Birmingham (1974).

3 Coltorti, M.; Simone, A. di; Cesare, D. di, and Rinaldi, M.: Effect of phenobarbital administration on liver and plasma gamma-glutamyl transpeptidase: clinical and experimental investigations. Acta vitamin. enzymol. *27:* 217–232 (1973).
4 Ideo, G.; Franchis, R. de; Minno, E. del, et Dioguardi, N.: Phenobarbitone increases rat liver gamma-glutamyl transpeptidase. Lancet *ii:* 825–826 (1971).
5 Orlowski, M. and Szewczuk, A.: Colorimetric determination of gamma-glutamyl transpeptidase activity in human serum and tissues with synthetic substrates. Acta biochim. pol. *8:* 189–200 (1961).
6 Rosalki, S.B.: Gamma-glutamyl transpeptidase. Adv. clin. Chem. *17:* 53–107 (1975).
7 Shaw, F.D.: The effect of mercuric chloride intoxication on urinary γ-glutamyl transpeptidase excretion in the sheep. Res. vet. Sci. *20:* 226–228 (1976).

J.P. Braun, VMD, Ecole Nationale Vétérinaire, Laboratoire Biochimie Médicale, *F-31076 Toulouse-Cedex* (France)

Enzymes in Health and Disease. Inaug. Scient. Meet. Int. Soc. Clin. Enzymol., London 1977, pp. 50–53 (Karger, Basel 1978)

Diagnostic Significance of Liver Cell Heterogeneity[1]

W. G. GUDER, H. SCHMID and U. SCHMIDT

Institute of Clinical Chemistry, Schwabing City Hospital, Munich;
Institute of Pathology, University of Tübingen, and Enzyme Laboratory, Medical Policlinic, University of Basel, Basel

Introduction

Liver cell heterogeneity is a well-known phenomenon in the histochemical literature [4, 5] but has until recently not been considered in biochemical studies on liver metabolism and clinical enzymology. Thus, the earlier observation of SHANK *et al.* [8] that alanine but not aspartate aminotransferase exhibits an activity gradient along the liver cell cords has been neglected in the interpretation of serum transaminase activity ratios found under various pathological conditions.

Recent studies in our laboratories confirmed and extended earlier findings in livers from rats and humans. The results obtained support the concept that intralobular localization of liver cell necrosis may influence serum enzyme activity ratios used in differential diagnosis of liver diseases.

Material and Methods

The following enzymes were determined in microdissected liver samples obtained from rats and humans: pyruvate kinase (EC 2.7.1.40) [2]; phosphoenolpyruvate carboxykinase (EC 4.1.1.32) [2]; fructose-1,6-bisphosphatase (EC 3.1.3.11) [7], and glutamate dehydrogenase (EC 1.4.1.3) [1]. Microdissection was performed according to LOWRY's techniques from lyophilized 16-μm liver slices. Enzyme analysis was performed with ultramicromethods on samples weighing 10–100 ng dry weight dissected from the portal and central parts of the liver lobule. Experimental details are given in the references cited above.

[1] Supported by the Deutsche Forschungsgemeinschaft (Sonderforschungsbereich 51, Munich, Project D 13 (GUDER), and Grant No. 217/6 (SCHMIDT).

Results

As observed earlier for transaminases and glutamate dehydrogenase in rat liver [8], portal and central enzyme activities of rate-limiting enzymes of carbohydrate metabolism were unevenly distributed over the liver lobule of the fed rat (table I). Glycolytic pyruvate kinase was found to be twice as high in central as in portal liver cells. In contrast gluconeogenic enzymes, phosphoenolpyruvate carboxykinase and fructose-1,6-bisphosphatase exhibited a reverse distribution pattern. Their activities decreased along the liver sinus with a 3 times higher activity in the portal region. The latter observation is quite in agreement with the distribution pattern of alanine aminotransferase found by SHANK *et al.* [8] and by WELSH [9].

Table I. Liver cell heterogeneity in the rat

Enzymes	Enzyme activities, U/g dry weight		
	portal	central	ratio
Pyruvate kinase	282^{+74}_{-18} (18)	644^{+72}_{-40} (18)	0.44
PEP-carboxykinase	720^{+500}_{-280} (14)	250^{+230}_{-150} (16)	2.88
Fructose-1,6-bisphosphatase	35 ± 5 (12)	10 ± 3.3 (12)	3.5

Enzyme activities from the periportal and pericentral region of the liver lobule from fed rats as mean ± 25-percentiles, numbers in parentheses. For experimental details, see GUDER and SCHMIDT [2] and SCHMIDT *et al.* [7].

Table II. Liver cell heterogeneity in man

Enzymes	Enzyme activities, U/g dry weight		
	portal	central	ratio
Glutamate dehydrogenase	191 (95)	325 (96)	0.59
Fructose-1,6-bisphosphatase	70 (5)	28.3 (5)	2.5

Enzyme activities were determined in microdissected liver cell groups from the periportal and pericentral region of lyophilized liver slices. The results are given as means with the number of samples in parentheses. For further details, see GUDER *et al.* [1] and SCHMIDT *et al.* [7].

Starvation of rats for 48 h led to an increase of gluconeogenic enzymes in both areas and a decrease of pyruvate kinase in the central region [2].

As can be seen in table II, the data thus far obtained in human liver were in good agreement with studies in rat livers. The higher glutamate dehydrogenase activity found in the central region was also reported for rat liver by SHANK *et al.* [8]. The fructose-1,6-bisphosphatase activities obtained in a liver specimen taken during laparotomy were quite similar to those found in starved rats [7]. MORRISON and BROCK [3] also found comparable enzyme gradients in rat and human liver when they studied the distribution of alcohol dehydrogenase, which was also found to be higher in the central liver cells.

Discussion

The results obtained clearly demonstrate that enzyme equipment of liver cells depends on the localization of the cell in the liver lobule. Cells from the periportal region contain more gluconeogenic enzymes, whereas enzymes like pyruvate kinase, alcohol dehydrogenase and glutamate dehydrogenase are found to be more active in the pericentral hepatocytes.

For clinical enzymology, the distribution of those enzymes used in the diagnosis of liver diseases is of considerable interest. The inverse distribution pattern of alanine aminotransferase and glutamate dehydrogenase in the liver lobule could be one of the factors causing different enzyme activity ratios in various liver diseases [6].

If liver cell necrosis would lead to an equivalent enzyme efflux of these enzymes into the blood stream, higher transaminase to glutamate dehydrogenase ratios would be expected in liver diseases destroying periportal cells. On the other hand, relatively higher glutamate dehydrogenase activities are to be expected in central liver cell necrosis. In fact, very high glutamate dehydrogenase activities have been observed in acute liver cell necrosis caused by acute right heart failure. In a study of 16 cases, glutamate dehydrogenase activity was 40 times higher in acute right heart failure than in cases of viral hepatitis exhibiting the same transaminase activities [1]. These large differences indicate that, in addition to liver cell heterogeneity, other factors are also involved in causing serum enzyme ratios. Among these, the different intracellular localization of enzymes, different biological half-life times of serum enzymes and the time course of development of liver cell necrosis have to be considered.

These factors, however, did not explain the relatively high serum activities of glutamate dehydrogenase in cholestatic and metastatic liver diseases observed by SCHMIDT and SCHMIDT [6]. The fact that both conditions are characterized by a preferential necrosis of liver cells from the pericentral region support our concept that serum enzyme patterns reflect the intralobular distribution of liver cell necrosis.

The high glutamate dehydrogenase activities observed in carbon tetrachloride poisoning can also be explained by the present findings.

In conclusion, liver cell heterogeneity has proved a fruitful new aspect for the interpretation of serum enzyme activity ratios observed in various liver diseases. In addition to factors known previously, liver cell heterogeneity seems to be involved in causing different alanine aminotransferase over glutamate dehydrogenase ratios in serum under different pathophysiological conditions.

References

1 GUDER, W.G.; HABICHT, A.; KLEISSL, J.; SCHMIDT, U., and WIELAND, O.H.: The diagnostic significance of liver cell inhomogeneity. J. clin. Chem. clin. Biochem. *13:* 311–318 (1975).

2 GUDER, W.G. and SCHMIDT, U.: The distribution of pyruvate kinase and phosphoenolpyruvate carboxykinase (GTP) in the liver lobule of fed and starved rats. Hoppe-Seyler's Z. physiol. Chem. *357:* 1793–1800 (1976).

3 MORRISON, G.R. and BROCK, F.E.: Quantitative measurement of alcohol dehydrogenase in the lobule of normal livers. J. Lab. clin. Med. *70:* 116–120 (1967).

4 NOVIKOFF, A.B.: Cell heterogeneity within the hepatic lobule of the rat (staining reactions). J. Histochem. Cytochem. *7:* 240–244 (1959).

5 PETTE, D. und BRANDAU, H.: Enzym-Histiogramme und Enzymaktivitätsmuster der Rattenleber. Enzymol. biol. clin. *6:* 79–122 (1966).

6 SCHMIDT, E. und SCHMIDT, F.W.: Methode und Wert der Bestimmung der Glutaminsäure-Dehydrogenase im Serum. Ein Beitrag zur Bedeutung der Untersuchung von Enzym-Relationen im Serum. Klin. Wschr. *40:* 962–969 (1962).

7 SCHMIDT, U.; SCHMID, H., and GUDER, W.G.: Liver cell heterogeneity: the distribution of fructose-bisphosphatase in fed and fasted rats and in man. Hoppe Seyler's Z. physiol. Chem. *359:* 193–198 (1978).

8 SHANK, R.E.; MORRISON, G.; CHENG, C.H.; KARL, I., and SCHWARZ, R.: Cell heterogeneity within the hepatic lobule (quantitative histochemistry). J. Histochem. Cytochem. *7:* 237–241 (1959).

9 WELSH, F.A.: Changes in distribution of enzymes within the liver lobule during adaptive increase. J. Histochem. Cytochem. *20:* 107–111 (1972).

W.G. GUDER, MD, Schwabing City Hospital, Institute of Clinical Chemistry, *D-8000 München* (FRG)

Enzymes in Health and Disease. Inaug. Scient. Meet. Int. Soc. Clin. Enzymol., London 1977, pp. 54–58 (Karger, Basel 1978)

Human Microsomal Carboxylesterase (EC 3.1.1.1.)

Distribution in Several Tissues and Some Preliminary Observations on Its Appearance in Serum

Wolfgang Junge[1]

Städtisches Krankenhaus Kiel, Zentrallabor, Kiel

Introduction

In 1974, we succeeded in purifying human liver carboxylesterase (EC 3.1.1.1) to homogeneity [3]. This enzyme is found in the microsomal fraction of a liver homogenate. Its molecular and kinetic properties have been determined [3, 4, 6]. In this report, results of investigations are described which were undertaken to study the following problems: First, does the carboxylesterase also occur in other organs than liver, or even in serum or other body fluids? Secondly, if it is normally not found in the bloodstream, does it appear in the serum under certain pathological conditions?

Material and Methods

Samples of human tissues were obtained at autopsy from the local Pathological Department. Homogenates were made by means of a blendor-type homogenizer in a 9-fold volume of ice-cold distilled water. Triton X-100 was then added (final concentration 0.5% v/v) to solubilize the carboxylesterase from the microsomal membranes. After 4 h incubation, the samples were centrifuged at 8,000 *g* and the supernatant was tested for esterase kinetically (substrate: 0.01 M methyl butyrate, pH 7.5, 30 °C using the pH-stat technique [3]) and immunologically by means of the radial immunodiffusion technique: samples are filled into wells cut in an agar gel which contains anti-carboxylesterase serum from rabbit [3]. After 48 h of incubation, the plates are washed, dried and stained for

[1] The author wishes to thank Miss B. Philipp for her expert technical assistance and Miss I. Falk, Miss S. Löhndorf and Mrs. H. Thiessen for collecting the serum samples.

esterase activity using α-naphthyl acetate as the substrate and Fast Red TR salt as the coupling reagent. For quantitative evaluation, each plate was standardized with solutions of highly purified liver carboxylesterase. All activities are expressed in U/l.

Results

The determination of carboxylesterase levels in the homogenates of various organs by a kinetic method is complicated by the fact that no specific substrate of this enzyme is known. Most, if not all, esters which are hydrolysed by carboxylesterase are also substrates of various other hydrolytic enzymes, e.g. lipase (EC 3.1.1.3), cholinesterase (EC 3.1.1.8), arylesterase (EC 3.1.1.2), acetylesterase (EC 3.1.1.6), lysophospholipase (EC 3.1.1.5) and several proteases.

The problem of this nonspecificity was circumvented by using an immunoenzymatic technique which allows the specific detection of carboxylesterase down to concentrations of 20 U/l. It should be emphasized that the specificity of this method is based on the immunological reaction and not on the enzymatic one.

The results of the kinetic and immunological determination of carboxylesterase levels in various tissues are compiled in table I. As expected,

Table I. Methyl butyrate hydrolysing activity and immunologically determined carboxylesterase levels in human tissues (autopsy material)

Source	n	Methyl butyrate (mU/g) mean ± SD	Immuno-diffusion (mU/g) mean ± SD
Liver	8	204,000 ± 92,000	192,000 ± 82,000
Lung	6	8,900 ± 3,400	8,300 ± 3,100
Spleen	5	3,900 ± 2,000	3,040 ± 1,500
Kidney	7	2,120 ± 700	980 ± 350
Pancreas	5	8,000 ± 2,700	0[1]
Thyroid	3	2,120 ± 2,070	1,960 ± 1,740
Testis	3	1,130 ± 400	1,030 ± 450
Adrenal gland	3	2,550 ± 780	1,850 ± 630
Heart	5	1,780 ± 780	1,780 ± 1,080
Muscle	3	670 ± 470	430 ± 40
Fat	3	4,400 ± 2,300	4,700 ± 3,300
Brain	3	670 ± 80	250 ± 170

[1] Except for one case, in which a weak precipitate was observed.

the activities are in general higher in the case of the kinetic experiments. The carboxylesterase was neither detectable in serum nor in any other body fluid.

These results indicate that the nonspecific carboxylesterase can be regarded as a highly specific liver enzyme. 1 g of liver contains about 1.2 mg of esterase as calculated from the immunological experiments. The high amount of carboxylesterase in this organ was the basis for investigation of whether or not the enzyme appears in the bloodstream under certain pathological conditions, particularly in those involving the liver. By means of the above described immunoenzymatic technique, more than 1,000 sera of patients who had elevated levels of enzymes routinely used for the diagnosis of primary or secondary liver diseases were examined for carboxylesterase. The enzyme was detected in about 30% of the cases. Figure 1 shows an experiment in which various positive sera have been quantitatively evaluated. In many cases, however, there was only a weak, faintly stained precipitate visible which did not allow exact quantitative determination. The activity ranged from ~20 to 6,800 U/l. Among the positive sera, 57 had activities higher than 200 U/l. These cases were differentiated according to their essential clinical and/or pathological diagnosis. The results are compiled in figure 2. The highest activities of carboxylesterase were found in diseases causing acute or subacute liver congestion: acute right heart or global cardiac failure, cardiogeneic shock, status asthmaticus or pulmonal artery embolism. A profound damage of liver cells with subsequent liberation of the microsomal bound esterase also occurs under various toxic conditions, e.g. alcohol or drug abuse, bile duct obstruction, and likewise in acute pancreatitis.

The possible presence of carboxylesterase in serum may lead to an interference with the determination of certain other hydrolytic enzymes. For example, the determination of arylesterase using β-naphthyl acetate [1] or the measurement of postheparin-lipase with monoolein [2] may give too high values because both compounds are also good substrates of human liver esterase [6]. A well-known problem which up to now is by no means satisfactorily resolved is the disturbance of the determination of serum lipase activity by carboxylesterases. Various compounds have been suggested as being specific substrates for lipase, e.g. short- or medium-chain mono- or triglycerides, Tweens or numerous chromophores esterified with long-chain fatty acids. However, all of these are also substrates of the carboxylesterases, as was shown in a detailed study by RICK [8], who used the purified enzyme from pig liver. Similar results were found in our study with the highly purified human liver carboxylesterase [6]. In addition, we could show that

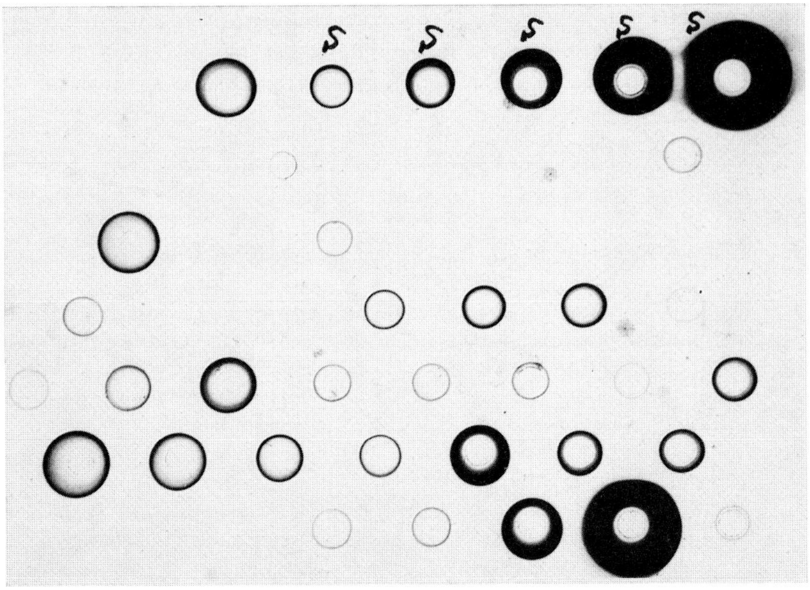

Fig. 1. Quantitative determination of carboxylesterase in 49 positive sera. S = Standards of purified liver carboxylesterase. Antiserum concentration, 0.025% (v/v).

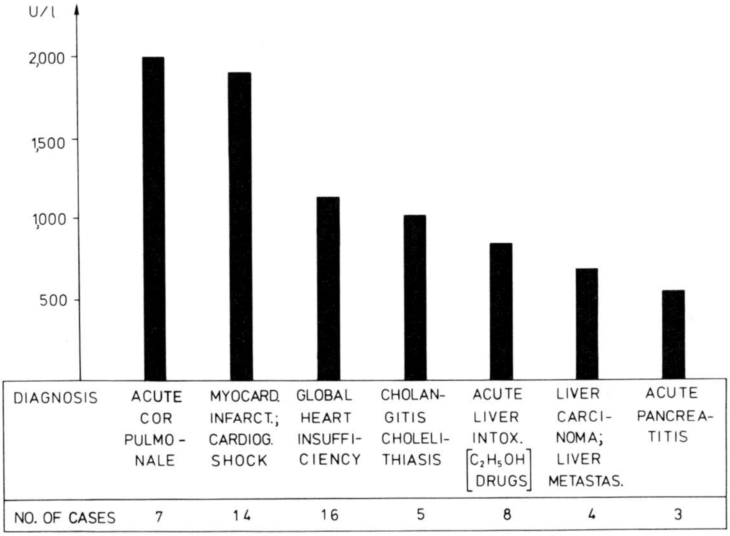

Fig. 2. Carboxylesterase activities as determined immunoenzymatically in the sera of 57 patients suffering from the indicated diseases. The main diagnosis was used only.

vinyl 8-phenyloctanoate, which was recently introduced as a highly specific substrate for serum lipase in routine diagnosis [7], is also hydrolysed with a considerable rate by human liver esterase [5].

References

1 BURLINA, A. and GALZIGNA, L.: A new and simple procedure for serum arylesterase. Clin. chim. Acta *39:* 255–257 (1972).

2 FRITSCH, W.-P. und RICK, W.: Postheparin-Lipase; in BERGMEYER Methoden der enzymatischen Analyse; 3. Aufl., Band I, pp. 854–861 (Verlag Chemie, Weinheim 1974).

3 JUNGE, W.; HEYMANN, E.; KRISCH, K., and HOLLANDT, H.: Human liver carboxylesterase. Purification and properties. Archs Biochem. Biophys. *165:* 749–763 (1974).

4 JUNGE, W. and KRISCH, K.: The carboxylesterases/amidases of mammalian liver and their possible significance. Crit. Rev. Toxicol. *3:* 371–434 (1975).

5 JUNGE, W. und LEYBOLD, K.: Hydrolyse des neu eingeführten Lipasesubstrats 8-Phenyloctansäurevinylester durch Carboxylesterase aus Menschenleber. Laboratoriumsmedizin *1:* 149–150 (1977).

6 JUNGE, W. and SCHLAAK, H.E.: Human liver carboxylesterase. Kinetic properties (in preparation).

7 MYRICK, J.E.: A spectrophotometric method for the kinetic measurement of lipase activity; dissertation, Birmingham (1976).

8 RICK, W.: Der Secretin-Pankreozymin-Test in der Diagnostik der Pankreasinsuffizienz. Internist *11:* 110–117 (1970).

W. JUNGE, MD, Städtisches Krankenhaus, Zentrallabor, Metzstrasse 53, *D-2300 Kiel* (FRG)

Enzymes in Health and Disease. Inaug. Scient. Meet. Int. Soc. Clin. Enzymol., London 1977, pp. 59–67 (Karger, Basel 1978)

Differential Diagnostic Value of Enzyme Histochemistry in Liver Biopsy Specimens

E. Kautzsch

Medical Department, Städtisches Krankenhaus München-Oberföhring, München

This paper records the data from 43,000 enzyme histochemical examinations from 4,500 human liver samples obtained by needle biopsy. Work was done up to now since 1962. Lactic acid dehydrogenase (LDH), succinic dehydrogenase (SuDH), glutamic acid dehydrogenase (GluDH), acid (a) and alkaline (alk) phosphatases (Ph) and adenosine-tri-phosphatase (ATPase) were stained and visualized [Kautzsch, 1967].

First of all, typical alterations were expected, but after 15 years we have learned:

(1) Typical enzyme responses to distinct biochemical or functional reactions of liver tissue are the exception rather than the rule.

(2) No defined enzyme pattern is attributable to any liver disease. Only in viral hepatitis is there a typically localized activity of alkPh in borderline structures of sinusoids and there is also a typical enzyme pattern of liver cell necrosis caused by viral injury [Kautzsch, 1973].

(3) Some general aspects have been found by histochemical staining of human liver enzymes: (a) An altered distribution of LDH, SuDH or GluDH indicates mitochondrial injury in single hepatocytes or in groups of liver cells. This may occur in chronic aggressive hepatitis with piecemeal necrosis, in toxic injured liver tissue or in histotoxic hypoxia. (b) Some alterations in mesenchymal cells such as Kupffer cells or macrophages depend on damaged lysosomes or dense bodies. They become obvious by acid phosphatase staining. (c) Raised activities of ATPase or alkPh in borderline structures are consistent with any developing process in liver disease. Augmented activity of the membrane-bound ATPases and raised activity of alkPh along the sinusoids is conspicuous (fig. 1). This is an important sign demonstrating activity of any liver disease. Early visualization studies by Danielli

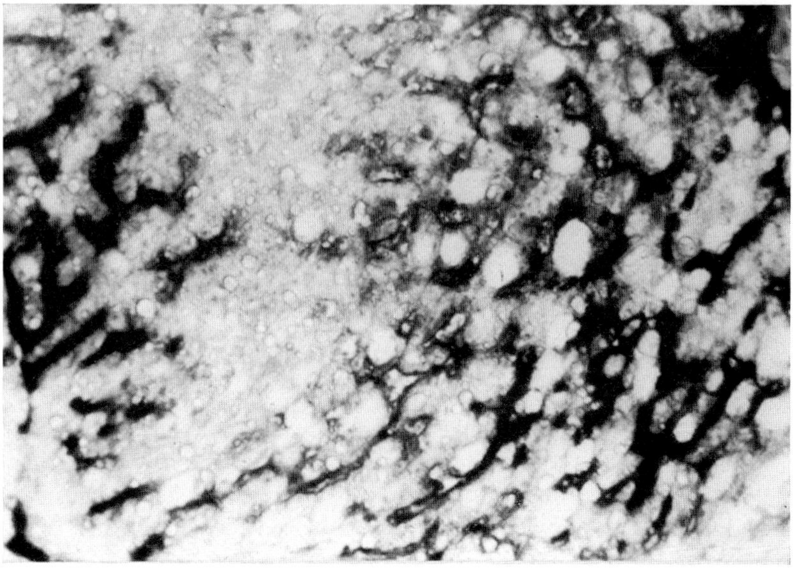

Fig. 1. Increased alkPh in sinusoidal borders (liver cirrhosis).

[1952] have shown that alkPh activity was high in areas of molecular transport. ATPase is an enzyme acting on acid anhydride bonds in phosphoryl-containing anhydrides. In all cases with an acute increase of developing collagen, the activity of membrane ATPase (3.6.1.3) was intensified, mostly in the sinusoid borders, in the membranes of bile canaliculi and in the membranes of blood vessels in the periportal fields. Protection against collagen may induce hydrolase activity to reject collagen in an early stage.

(4) An alteration in enzyme activity and localization in the liver tissue does not depend solely on a disease localized in the liver but may occur in other diseases. Mostly Kupffer cells are concerned.

(5) In diseases of the liver itself, the changing of enzyme pattern normally takes place in the whole liver and not only locally.

(6) There is no correlation between enzyme activity in liver tissue and in blood serum, particularly ATPase, alkPh and aPh.

Some examples may explain and demonstrate those facts mentioned above, in regard to differential diagnosis.

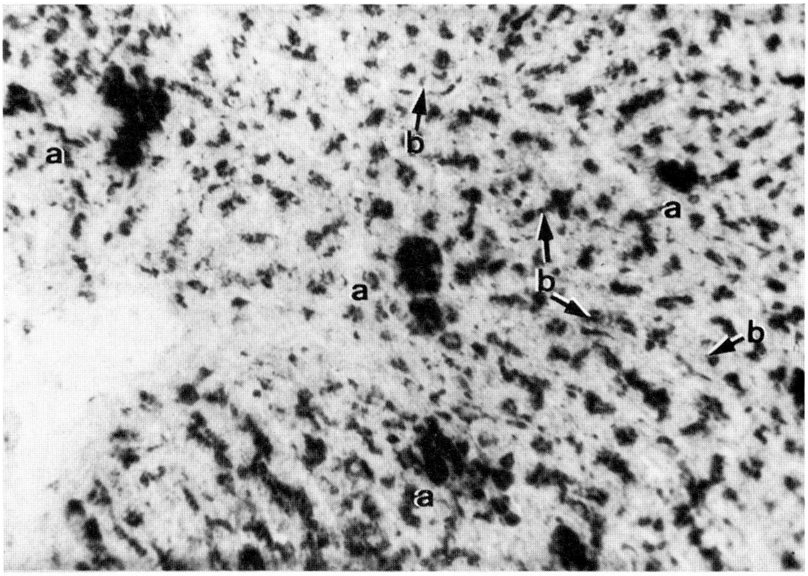

Fig. 2. Intensified aPh in spotty necrosis (a) and some Kupffer cells (b) (viral hepatitis).

A. Viral Hepatitis

Hepatocytes attacked by virus are but slowly killed. In the first stage microsomes are damaged. This stage is seen well by electron microscopy. In the second stage, the activity of mitochondral enzymes such as SuDH and GluDH decrease. At the same time, there is an augmentation of acid phosphatase in the lysosomes. In the third stage, lysosomes are damaged and self-digestion follows. This feature is characterized by 'spotty necrosis', sometimes attended by stainable diffuse iron and ceroid pigment (fig. 2, 3) [Kautzsch, 1973].

This impressive feature is only to be seen in human viral hepatitis or in experimental murine viral hepatitis [Allison and Sandelin, 1963]. The raised activity of aPh in spotty necrosis of human viral hepatitis persists for about 3 weeks in uncomplicated cases. At first, the spotty necrosis is spread near the periportal fields, later near the central vein of the liver lobule. Further viral infection results in the reappearance of spotty necrosis near

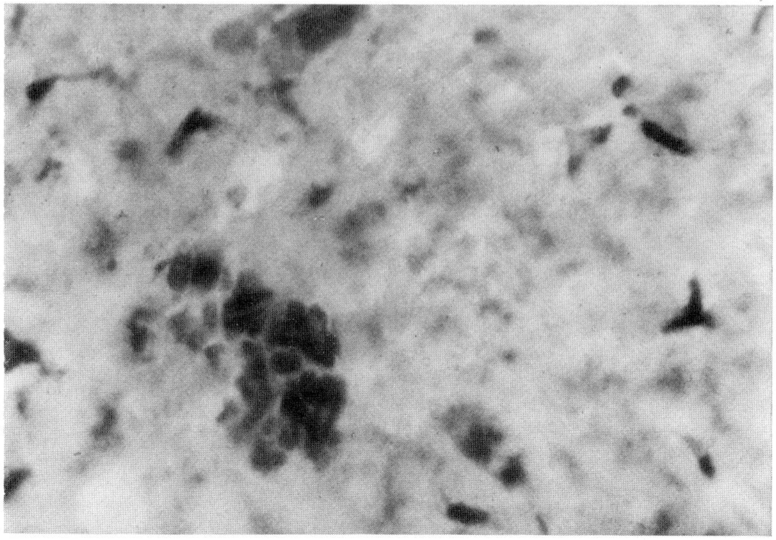

Fig. 3. Increased aPh activity in one big spotty necrosis surrounded by Kupffer cells with increased activity of aPh (viral hepatitis).

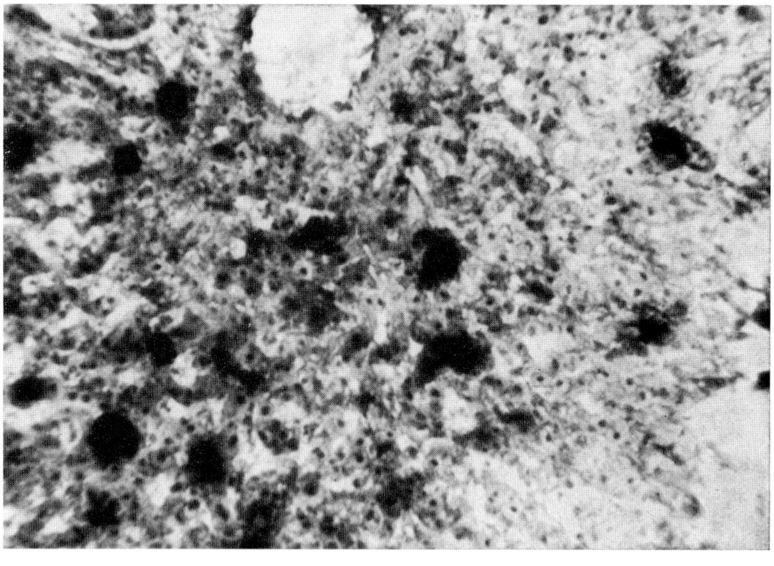

Fig. 4. ATPase increased in spotty necrosis (viral hepatitis).

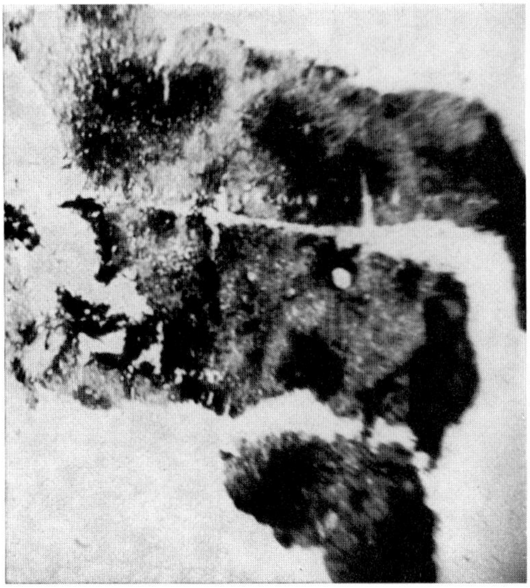

Fig. 5. Cumulating intensified activity of alkPh (viral hepatitis).

the periportal fields. Otherwise, the augmented activity of aPh disappears and self-digestion seems to be completed. But now, there is an increase of ATPase activity coming from Kupffer cells and macrophages which have assembled round the spotty necrosis in the meantime. This fourth stage in the removal of damaged liver cells lasts about 100 days and is still obvious when all other signs of viral hepatitis have already disappeared (fig. 4).

Another characteristic sign of acute viral hepatitis in man is demonstrated by intensified staining of alkPh in the sinusoidal borders as in any other acute liver disease (fig. 5). Only the main locality around the periportal fields and central veins is a certain sign of continuous acute viral hepatitis, particularly if 'aPh-stained' spotty necrosis comes together with augmented alkPh activity in the sinusoidal borders.

B. Cirrhosis of the Liver

Intensified ATPase activity in bile ducts of the periportal fields is supposed to be a sign of primary biliary cirrhosis (PBC), even when asymptomatic (fig. 6). Increased activity of ATPase in canaliculi (fig. 7) may also suggest PBC. This is not found in cholestasis due to causes other than

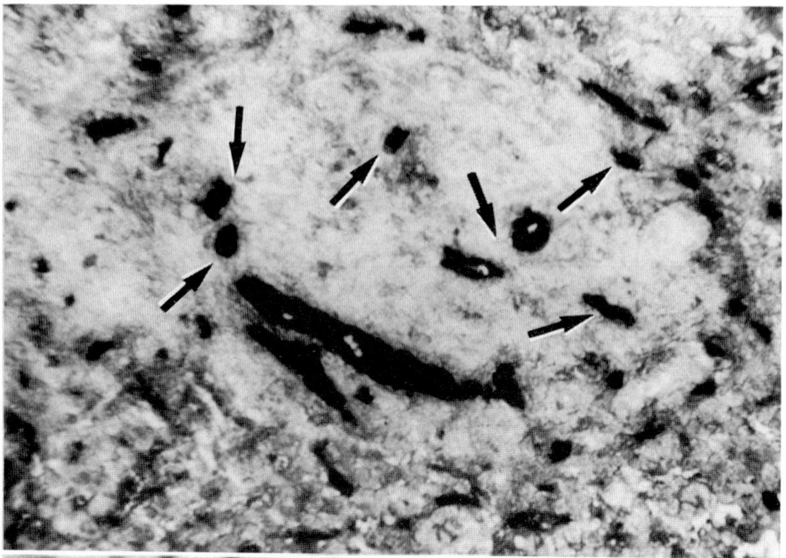

6

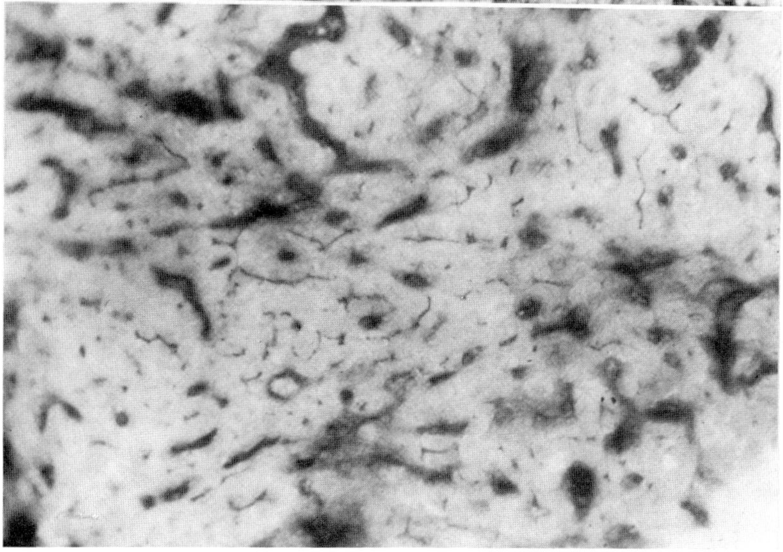

7

Fig. 6. Intensified staining of ATPase in periportal ductules (arrows) (primary biliary cirrhosis).

Fig. 7. Augmented ATPase activity in canaliculi (primary biliary cirrhosis).

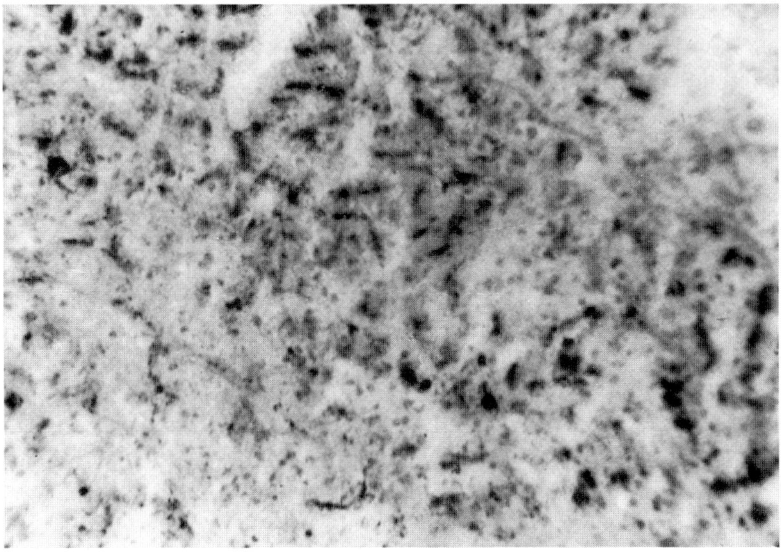

Fig. 8. Different diminished activity of aPh (liver cirrhosis).

cirrhosis. An augmented activity of ATPase both in the bile ducts and canaliculi together with a raised amount of alkPh in blood serum confirms the diagnosis of PBC.

Without stained canaliculi and without augmented staining of the ductules but with increased activity of alkPh in sinusoidal borders, portal cirrhosis of the liver is probable (fig. 1). A normal distribution and activity of both aPh and alkPh in these cases suggests an inactive cirrhosis while in an active cirrhosis the lysosomal activity of aPh diminishes in several parts of the lobule (fig. 8); thus, a different staining of lysosomal pericanalicular aPh indicates an active stage of cirrhosis.

C. Reactive Hepatitis

As already mentioned above, the activity of alkPh demonstrates generally whether or not active liver disease is to be expected. Especially in reactive hepatits (RH) this proves a very useful sign for prognosis. The changeover from RH to liver cirrhosis is not yet exactly known and remains a problem. Help may be obtained by study of alkPh. Kupffer cells in hepatic diseases show an augmented activity of aPh if the mesenchymal tissue is highly active, independent of the origin of the disease. Intensifield deposits of stainable iron in Kupffer cells may be present for a shorter time, therefore are less predictable. Cases without typical histologic alteration of liver tissue slices but with an increased activity of alkPh in sinusoids and with an increased activity of aPh in Kupffer cells should be diagnosed as reactive hepatitis.

D. Conclusion

Routine staining and visualizing marker enzymes such as both alkPh and aPh besides ATPase in human liver biopsy specimens indicate the activity of liver diseases, the stimulation of the Kupffer cells, and the course of viral hepatitis as well as of liver cirrhosis. The histochemical behaviour of the enzymes points to some rules which may be of differential diagnostic and of prognostic value.

Summary

By histochemical demonstration of 'marker enzymes' in liver biopsy slices, the phosphatases are noted to be of the highest differential diagnostic value. This assertion stems from the evaluation of 4,500 different human liver biopsies. In all cases of liver diseases, there is an increase of the activity of alkaline phosphatase in the sinusoid borders during progress of the disease. Increasing activity is spread through the whole liver tissue even if the real process of the disease is located only in one part of the liver. Increased ATPase activity precedes the development of collagen. The activity of acid phosphatase, a lysosomal marker enzyme, is diminished in cirrhosis of the liver and increased in Kupffer cells after activation of the mesenchyme as in viral hepatitis or in any infectious diseases outside the liver. In this regard, the activity of acid phosphatase points exactly to the resolution of acute viral hepatitis. The behaviour of alkaline phosphatase is a reliable parameter which helps to complete the usual histological examination.

References

ALLISON, A.C. and SANDELIN, K.: Activation of lysosomal enzymes in virus infected cells and its possible relationship to cytopathic effects. J. exp. Med. *117:* 879–887 (1963).

DANIELLI, J.F.: Structural factors in cell permeability and secretion. Symp. Soc. exp. Biol. *6:* 1–15 (1952).

KAUTZSCH, E.: Activity and localization of liver enzymes in acute and chronic liver diseases. Liver research. 3rd Int. Symp. IASL, Tokyo. Tijdschr. Gastroent. *1967:* 194–205.
KAUTZSCH, E.: Alteration of enzyme histochemistry in liver cells attacked by human hepatitis virus. Acta hepato-gastroent. *20:* 303–306 (1973).

E. KAUTZSCH, MD, Städtisches Krankenhaus München-Oberföhring, Medizinische Abteilung, Oberföhringer Strasse 156, *D-8000 München 81* (FRG)

Enzymes in Health and Disease. Inaug. Scient. Meet. Int. Soc. Clin. Enzymol., London 1977, pp. 68–72 (Karger, Basel 1978)

Isoenzyme Patterns of *N*-Acetyl-β-Glucosaminidase from Healthy and Diseased Human Livers[1]

G. Pott, N. van Husen and U. Gerlach

Medical Clinic and Policlinic, Division of Gastroenterology,
University of Münster, Münster

Introduction

Studies on *N*-acetyl-β-glucosaminidase (EC 3.2.1.30), a lysosomal glycosidase, involved in the breakdown of glycosaminoglycans, showed increase of this enzyme in sera and liver tissues from patients with chronic active liver diseases (CALD) [3, 5–7]. As described in previous publications, isoenzymes of β-glucosaminidase can be separated by ion-exchange chromatography [9], electrophoresis [9] and electrofocusing [2, 8, 10]. Our studies are performed to solve the question whether there is a change in the isoenzyme patterns of β-glucosaminidase in liver tissues from patients with CALD or not. Furthermore, isoenzyme patterns are compared to the activity of total β-glucosaminidase, collagen content and histologically estimated activity of liver fibrosis.

Materials and Methods[2]

Liver tissues from healthy persons (n = 7) and from patients with CALD (n = 20) were taken by autopsy and stored at –20 °C for up to 2 months.

Homogenization of liver tissue and *enzyme assay* of β-glucosaminidase, using a *p*-nitrophenyl-*N*-acetyl-β-glucosaminide at pH 4.4, were done according to previous publications [1, 7, 8]. *Isoenzyme bands* of β-glucosaminidase were separated by electro-

[1] Supported by a grant (Ge 251/5–SFB 104) from the Deutsche Forschungsgemeinschaft, Bonn-Bad Godesberg.

[2] With skilful technical assistance of Mrs. D. Chi-Boesler.

focusing in thin layers of polyacrylamide (Multiphor, LKB, Bromma, Sweden; 5% gel; 1% ampholytes) [8]. Activity of enzyme bands was visualized using a naphthyl-*N*-acetyl-β-glucosaminide. Gels were quantitated by densitometry. pH gradients were measured by a surface pH electrode (Ingold, Zürich). Collagen was estimated as described previously [12]. Activity of total β-glucosaminidase is given in U/g protein.

According to published results and criteria [7], *activity* of *fibrosis* in the liver was determined by histological examination of the liver tissue (+ to + + +).

Results and Discussion

Isoenzyme patterns of β-glucosaminidase are demonstrated in figure 1. Measurements of pH gradients by a surface pH electrode showed isoenzyme A at pH 5.0 and B at pH 7.3, as described before [8, 10]. Between pH 5.5 and 7.0 enzyme bands are visible, whose identification is not clear. They are named as P bands [13].

In *patients with CALD*, isoenzyme A increases corresponding to total activity of β-glucosaminidase and histologically estimated activity of fibrosis in the livers (table I). As demonstrated earlier by Dorfman and co-workers [11], using cultured fibroblasts from patients with β-glucosaminidase deficiency (Sandhoff disease), isoenzyme A is more specific against hexos-

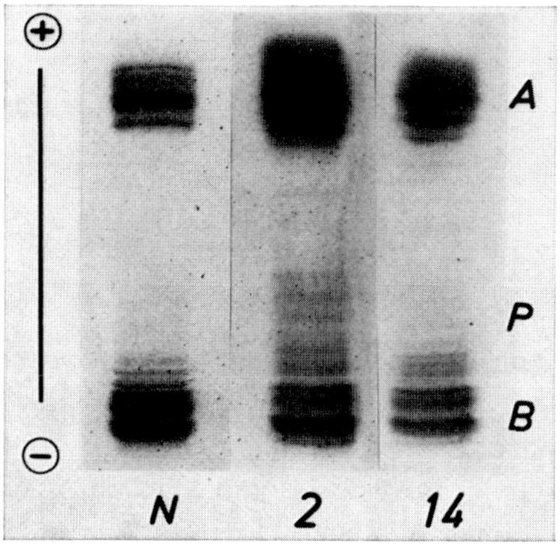

Fig. 1. Isoenzyme patterns of *n*-Acetyl-β-glucosaminidase from liver tissue. N = Normal liver tissue. For details see No. 2 and 14 of table I. Isoenzyme separation was done by electrofocusing in thin layers of polyacrylamide gel: 5% gel; 1% ampholytes (Multiphor, LKB).

Table I. Activity of β-glucosaminidase, isoenzyme A of β-glucosaminidase and fibrosis in liver tissue, taken by autopsy from normal persons (n = 7) and patients with CALD (n = 20). In comparison, collagen content of liver tissue is given. Definition of diagnosis according to LEEVY and TYGSTRUP [4]

No.	Diagnosis	Activity of fibrosis	Isoenzyme A/B+P	Activity of β-NAG, U/g protein	Collagen mg/g fresh liver tissue
N	normal liver (n = 7)	∅	40/60	30–50	3.5–8.5
Alcoholic liver disease					
1	hepatitis	+++	75/25	84.2	16.4
2	hepatitis	+++	70/30	79.5	22.8
3	cirrhosis	+++	65/35	94.8	54.5
4	cirrhosis	+++	65/35	78.3	39.5
5	cirrhosis	++	55/45	60.1	41.0
6	cirrhosis	++	50/50	62.8	35.7
7	hepatitis	++	55/45	58.7	13.8
8	cirrhosis	+	50/50	54.0	45.7
9	fatty liver	(+)	60/40	45.7	8.7
10	fatty liver	(+)	60/40	39.7	8.5
Fibrosis and cirrhosis, following viral hepatitis					
11	cirrhosis	+++	65/35	69.0	28.7
12	cirrhosis	++	60/40	75.0	42.5
13	cirrhosis	++	60/40	68.0	47.2
14	cirrhosis	+	45/55	39.2	38.4
15	cirrhosis	+	45/55	42.8	35.7
16	biliary cirrhosis	+++	65/35	88.4	38.8
17	biliary cirrhosis	++	60/40	82.9	44.5
18	congestive cirrhosis	+	50/50	34.8	28.5
19	congestive fibrosis	+	50/50	38.5	9.2
20	congestive fibrosis	(+)	50/50	42.4	17.2

amines in glycosaminoglycans than isoenzyme B. Glycosaminoglycans and collagen are the main components in the connective tissue of liver fibrosis.

As described previously by many authors, both pathologists and biochemists, increase of collagen content of liver tissue indicates fibrosis in the liver. As seen in table I, no correlation is found between collagen content and activity of β-NAG, β-NAG A and of fibrosis in the liver. On the other hand, correlation is seen between collagen content and cirrhosis, active or inactive.

From these findings it seems that: (1) Isoenzyme A is responsible for the increase of total activity of β-glucosaminidase in the fibrotic livers of patients with chronic liver diseases. (2) Furthermore, elevation of β-glucosaminidase in sera of those patients is caused by isoenzyme A. (3) Isoenzyme A and total β-glucosaminidase activity increase corresponding to the activity of liver fibrosis. (4) In contrast, collagen content results from increased metabolism of connective tissue, remaining for a long time after fibrotic activity has decreased.

Summary

The activity of *N*-acetyl-β-glucosaminidase was demonstrated in healthy and diseased human livers. Due to the activity of fibrosis in chronic active liver diseases (CALD), an elevation of activity of isoenzyme A was observed. No correlation was found between activity of this isoenzyme and collagen content of the liver.

References

1 Findlay, J.; Levvy, G.A., and Marsh, C.: Inhibition of glycosidases by aldonolactones of corresponding configuration. Biochem. J. *69:* 467–476 (1958).

2 Hayase, K. and Kritchevsky, D.: Separation and comparison of isoenzymes of *N*-acetyl-β-D-hexosaminidase of pregnancy serum by polyacrylamide gel electrofocusing. Clin. chim. Acta *46:* 455–464 (1973).

3 Koizumi, T.; Mitsutani, N.; Kawata, H.; Wasa, M., and Yoshida, T.: *N*-Acetyl-β-hexosaminidase activity in hepatic fibrosis. Lab. Invest. *13:* 752–758 (1964).

4 Leevy, C.M. and Tygstrup, N.: Standardisation of nomenclature, diagnostic criteria and diagnostic methodology for diseases of the liver and biliary tract (Karger, Basel, 1976).

5 Nakamura, N.; Iwabori, N., and Koizume, T.: Serum hyaluronidase activity in chronic liver damage. Clin. chim. Acta *27:* 47–52 (1970).

6 Platt, D. und Leineweber, B.: Aktivitätsmessung lysosomaler und cytoplasmatischer Enzyme im menschlichen Serum bei Leberkrankheiten. Verh. dt. Ges. inn. Med. *75:* 352–354 (1969).

7 Pott, G.; Eberhardt, G. und Gerlach, U.: Zur Beurteilung der Mesenchymreaktion bei Patienten mit chronischen Leberkrankheiten. Z. Gastroent. *16:* 32–37 (1978).

8 Pott, G.; Chi-Boesler, D., and Gerlach, U.: Isoenzymes of *N*-acetyl-β-glucosaminidase from human liver and serum. Separation by electrofocusing in thin layers of polyacrylamide gel. J. clin. Chem. clin. Biochem. *16:* 15–18 (1978).

9 Robinson, D. and Stirling, J.L.: *N*-Acetyl-β-glucosaminidases in human spleen. Biochem. J. *107:* 321–327 (1968).

10 Sandhoff, K.: Auftrennung der Säuger *N*-Acetyl-β-D-hexosaminidase in multiple Formen durch Elektrofokussierung. Hoppe-Seylers Z. physiol. Chem. *349:* 1095–1098 (1968).

11 Thompson, J.N.; Stoolmiller, A.C.; Matalon, R., and Dorfman, A.: *N*-Acetyl-β-hexosaminidase: role in the degradation of glycosaminoglycans. Science *181:* 866–867 (1973).

12 Stegmann, H.: Determination of hydroxyproline. Hoppe-Seylers Z. physiol. Chem. *187:* 41 (1958).

13 Stirling, J.L.: Separation and characterisation of *N*-acetyl-β-glucosaminidases A and P from maternal serum. Biochem. biophys. Acta *271:* 154–162 (1972).

G. Pott, MD, Medical Clinic and Policlinic, Division of Gastroenterology, University of Münster, Westring 3, *D-44 Münster* (FRG)

Enzymes in Health and Disease. Inaug. Scient. Meet. Int. Soc. Clin. Enzymol., London 1977, pp. 73–86 (Karger, Basel 1978)

Developments in Methodology of Enzyme Assays

HANS ULRICH BERGMEYER

Boehringer Mannheim Research Center, Tutzing

Introduction

The topic of my lecture is *Developments in Methodology of Enzyme Assays*. I shall restrict myself to a discussion of measurements of the catalytic activity of enzymes since clinical enzymology, as far as methodology is concerned, is related mainly to this type of measurement. However, I cannot and I will not describe former developments up to now (though this itself would be very interesting as BUETTNER [5] showed with his article in *Medical World* about the history of clinical chemistry).

My introductory questions to be answered will be first: 'Where do we stand?', then 'What is to be measured?', and 'Where should it be measured?'

The rising costs of health care have given prime importance to those enzyme assays with high clinical relevance. And only these assays are the objects of national and international standardization efforts, because of their clinical value.

A number of European societies of clinical chemistry have already made recommendations while other societies, including American ones, are working on standardization of the most important enzyme assays. According to my knowledge, the following survey can be given (table I).

It is interesting that γ-GT has only just been standardized by the Scandinavian society, not yet by others. But the Expert Panel on Enzymes of IFCC is working on γ-GT now, too.

Regarding methodology, one is tempted to say that the chemistry of enzyme assays is satisfactory or will be satisfactory, in the laboratories of most countries all over the world within a short period of time. But, so far, the methods and procedures of preparation of the patient and the procure-

Table I. Standardized methods in clinical chemistry

Parameter	Standardized by			
AST	D	Sc	F	IFCC
ALT	D	Sc	F	IFCC[1]
CK	D	Sc	F	
LDH	D	Sc		
HBDH	D			
AP	D	Sc	F	
LAP	D			
GLDH	D			

D = German, Sc = Scandinavian, F = French societies.
[1] In press.

Table II. Relationship between age and body weight of men and women. AST in U/*l* according to [13].

Age years	Men			Women		
	relative body weight		mean	relative body weight		mean
	⩽1.2	>1.2		⩽1.2	>1.2	
⩽30	10.2	14.1	10.2	9.7	–	9.7
31–40	10.4	12.0	11.1	8.5	8.6	8.6
41–50	11.1	11.9	11.7	8.5	9.3	8.9
51–60	10.3	11.1	11.0	9.8	9.5	9.6
>60	10.9	10.8	10.9	9.5	10.0	9.8
Mean	10.8	11.5	11.1	9.0	9.4	9.2
n	395	313	708	265	403	668

ment and pretreatment of the sample before it reaches the laboratory are not satisfactory. Here, a lot of standardization work still has to be done.

Procedures involving the patient include the way in which samples are drawn (e.g. with or without venestasis). Furthermore, the state of the patient is of greatest importance. It is well known that the state of nutrition plays an important role if glucose, fatty acids or amino acids are to be determined. Also, it is nonsense to compare hormone values of samples drawn at day or at night time. At the very beginning of method standardization for enzyme activities, LAUDAHN *et al.* [9a] and others [13] drew attention to the fact that aminotransferase activities depend on sex, age and body weight, and vary with daily rhythms (table II).

For the discussion of developments in methodology which follows, we will refer only to the methods applied to the sample.

'What to Measure?'

We can measure enzymes and iso-enzymes. The advantage of the measurement of iso-enzymes is well known in the case of LDH and more recently for other enzymes. The determination of the CK iso-enzymes especially has become more and more important for the diagnosis of myocardial infarction.

'Where to Measure?'

Enzymes or iso-enzymes can be determined in principle in all body fluids, in all organs and cells. For clinical chemistry, routine laboratory enzyme activities are mainly measured in blood, plasma or serum, urine, and biopsy material.

Perhaps attempts to measure enzymes in urine, such as myokinase after myocardial infarction, will be more successful. However, measurements in biopsy material will be limited to special laboratories. Since the catalytic activity of enzymes in clinical routine laboratories is measured mainly in serum or plasma, let us limit our further discussion to these samples.

Developments in Methodology of Enzyme Assays

For the determination of the catalytic activity or catalytic activity concentration of an enzyme quantitatively it is necessary to have: (1) a method; (2) reagents; (3) equipment, and (4) personnel (fig. 1).

There is a close interdependency between method and equipment, between both and well-trained personnel performing the assay, and last but not least between method and reagents. In this lecture, I would like to stress only the point 'method'.

A method in clinical chemistry has to deliver data as an aid for medical diagnosis. These data can be seen under two points of view: (1) data for the establishment of a diagnostic hypothesis (these data have a value like a symptom), and (2) data for a yes or no answer on a diagnostic hypothesis.

One has to differentiate carefully between these two functions of

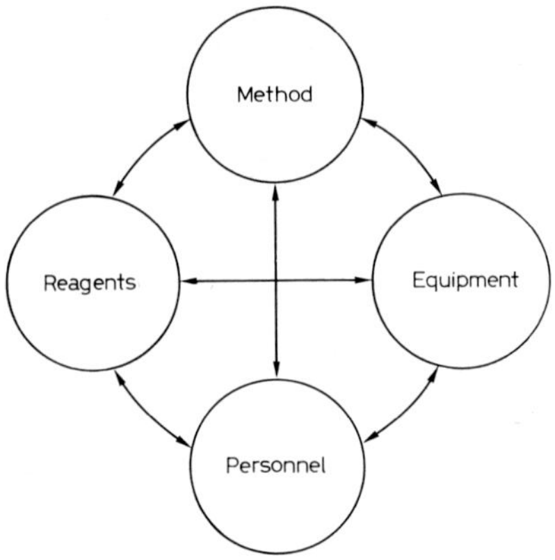

Fig. 1. Interdependencies of requirements for the determination of a catalytic activity of an enzyme.

laboratory data: In the second case, where the diagnostic hypothesis already exists, a few parameters with limited precision and accuracy will be enough. However, in their function as a symptom, a broader spectrum of laboratory data is necessary; their precision and accuracy must be high because differences and ratios often are used.

Since the second kind of laboratory data – also called 'real time analysis' – at present are restricted to the determination of substance concentrations on test strips, and do not extend to the determination of the catalytic activity of enzymes, in the following we need to consider only methods with high demands on precision and accuracy, mainly used in the central laboratory.

There are two possibilities for enzyme determinations: (1) based on catalytic activity (classically), and (2) based on protein measurement (immunologically).

The latter principle, advocated by our colleague MARC ROTH, is in progress; nevertheless, the first is the basis of the overwhelming majority of routine determinations.

Thus, considering methods for the determination of the catalytic activity of enzymes I would like to stress three points: (1) new measurement principles; (2) improvements of methods, and (3) advances in theoretical treatment of enzyme reactions.

Measurement Principles

Among the general measurement principles, only a few can be considered new in their application to clinical chemistry.

Measurement principles:

Photometry	Microcalorimetry and use of thermistors
Fluorimetry	Electrochemical methods
Nephelometry	(conductometry, electrodes)
Luminescence methods	Methods with carrier-fixed enzymes
Radioisotope methods	

We all know that photometry is the most widely used measurement principle in the routine clinical laboratory. Only in some cases fluorimetry is applied and nephelometry (also laser nephelometry) has now been introduced in a few cases, e.g. for the determination of human lysozyme and lipase. Whilst radioisotope methods have their limited place in enzyme analysis, luminescence methods are not so widely used for the routine determination of catalytic activities of enzymes.

The measurement principles mentioned in the second block are mainly used for the enzymatic determination of metabolites; they are under development but need not be considered here for the determination of the catalytic activity of enzymes.

Improvement of Methods

Development in methodology includes general improvement of existing methods and methods for new serum parameters.

Let me summarize first briefly the most important general improvements of methods:

First, we have the redetermination [1, 15] of absorption coefficients of NAD(P)H, of *p*-nitrophenol and others for which we will have lectures at this congress. I think it is very important to have accurate values now for the calculation of our basic parameters.

Second, we made the important γ-GT determination more convenient by introducing a new soluble substrate [11a, b].

Third, we have made CK determination more reliable by introducing a new activator, other than glutathione [12a, b]. Furthermore, there was worked out a convenient procedure for the determination of the CK-MB iso-enzyme by immunological inhibition of the M subunit [10].

Fourth, we optimized the important determination of AST and ALT using pyridoxal phosphate as coenzyme [2, 3]; both methods are going to be standardized by IFCC in the immediate future [7].

Fifth, we see advances in amylase determination, e.g. using short-chain polysaccharides [8].

Sixth, clinical chemistry now uses new substrates for color tests in the determination of blood-clotting proteases [14].

I think that this is a proud record of success in development of methods in recent years. More fundamental is the development of new diagnostic parameters. Here, the most exciting progress that I see is in the elucidation of the enzymopathy of porphyrin synthesis seen in acute hepatic porphyria; the measurement of the key enzyme uroporphyrinogen synthase in erythrocytes can be a routine method for the detection and classification of acute intermittent porphyria [6]. A lot of further work has been done in the field of lipid metabolism and its enzymes. But no routine method with high clinical significance could be made available up to now.

Advances in Theoretical Treatment of Enzyme Reactions

A modern method must be reliable, practicable and economic, with well-defined standards of accuracy and precision of the results. The definition of the terms 'precision' and 'accuracy' is well known and what has to be understood under 'practicability' is everything which leads to more convenience in practice. To fulfil these modern high demands without loss in accuracy and precision, it is necessary to know exactly what we do when we select or design a method. Modern knowledge in biochemistry makes it easier to illuminate the theoretical background of enzyme reactions. Thus, we can get more clarity concerning enzyme reactions, especially coupled enzyme reactions.

Because of the fact that mostly coenzyme-dependent reactions are used, the theory of reaction kinetics must involve reactions with two substrates (the coenzyme is one of them). In many cases, also, it must be considered that the primary reaction is coupled with an indicator reaction.

Primary reaction	$S_1 + S_2 \rightarrow P_1 + P_2$,
Indicator reaction	$P_1 + S_3 \rightarrow P_3 + P_4$.

In the case of NADH-dependent indicator reactions, S_3 is NADH.

Both reactions in general can be described by relatively simple rate laws. This is because no ternary enzyme substrate complex is formed in reactions according to the ping-pong bi-bi mechanisms (e.g. in case of AST and ALT).

On the other hand, we know that substrate and product inhibition may occur which must be considered.

Let me demonstrate recent progress in theoretical treatment of enzyme reactions, with the transferases AST and ALT as examples [2–4].

Primary Reactions

The aim all the time is to optimize enzyme reactions. To find out optimum conditions for measurement it is necessary at first to assess the following parameters: kind of buffer, ionic strength, pH and activators. Subsequently, under the conditions of these basic parameters, one should determine the absolute values of K_m in order to be able to calculate the optimum substrate concentrations.

In our case, we have two-substrate enzyme reactions, following the ping-pong bi-bi mechanism. The rate law is:

$$v_i = \frac{V}{1 + \frac{(K_m)_{S_1}}{[S_1]} + \frac{(K_m)_{S_2}}{[S_2]}} . \tag{1}$$

According to this equation, optimum reaction rate V, that means $v_i = V$, is only measured if the denominator of the right side of the equation becomes 1. That only is possible if $[S_1] \gg (K_m)_{S_1}$ and $[S_2] \gg (K_m)_{S_2}$ so that the second and the third term in the denominator becomes zero.

This cannot be achieved since, e.g. in the case of alanine aminotransferase, S_1 means L-alanine, whose solubility is limited to approximately 900 mmol/l; but, to make $(K_m)_{S_1}/[S_1] = O$, at least 2,000 mmol/l would be necessary. Therefore, maximum velocity cannot be measured due to technical reasons. One has to tolerate a certain deviation of v_i from V.

A similar situation for S_2 (2-oxoglutarate) exists. One cannot increase 2-oxoglutarate concentration indefinitely because of its light absorbance at the measuring wavelengths.

The first point we have learnt is that v_i cannot be equal to V.

Knowing this fact, we consider the relationship of the two substrate concentrations.

From equation 1 it follows that:

$$\frac{V}{v_i} - 1 = \frac{V - v_i}{v_i} = \frac{(K_m)_{S_1}}{[S_1]} + \frac{(K_m)_{S_2}}{[S_2]} = \frac{1}{F} ,$$

from this:

$$[S_1] = \frac{(K_m)_{S_1}}{\frac{1}{F} - \frac{(K_m)_{S_2}}{[S_2]}};$$

$$[S_2] = \frac{(K_m)_{S_2}}{\frac{1}{F} - \frac{(K_m)_{S_1}}{[S_1]}}.$$

Using the fraction v_i over V, represented by the factor F, we get from our primary reaction an equation showing the relationship between the concentrations of the two substrates S_1 and S_2. There is a hyperbolic relationship between the two substrates as has already been pointed out by the group of GARFINKEL [9b, c].

The second point that we have learnt is that for two-substrate enzyme reactions, there exist within certain experimental limits an unlimited number of substrate concentration pairs, situated on hyperbolas for a certain value of F, that will each give the same value of v_i as a fraction of V.

Figure 2 shows a family of hyperbolas valid for alanine aminotransferase. The concentration pairs of alanine (S_1) and 2-oxoglutarate (S_2) situated on the most curved parts are: the most economical pairs (the sum $S_1 + S_2$ is the lowest here) and of lowest variability (variation of S_1 yields lowest variation of S_2 and vice versa).

These clear relationships can be obscured by the fact that substrate or product inhibition occurs. In the case of aminotransferases measured with an indicator reaction, from the variety of inhibitions by substrates and products on the pyridoxal phosphate form and the pyridoxamine phosphate form of the enzyme which have been described, only substrate inhibition by 2-oxoglutarate must be considered in practice. The product inhibition by glutamate and by the keto acid is effective only up to a few percent and can be kept small under experimental conditions.

The substrate inhibition by 2-oxoglutarate is competitive; it is expressed by the inhibitor term $(1 + [I]/K_I)$ which is associated with K_m. Therefore, we get:

$$[S_1] = \frac{(K_m)_{S_1} \times \left(1 + \frac{[I]}{K_I}\right)}{\frac{1}{F} - \frac{(K_{mS_2})}{[S_2]}},$$

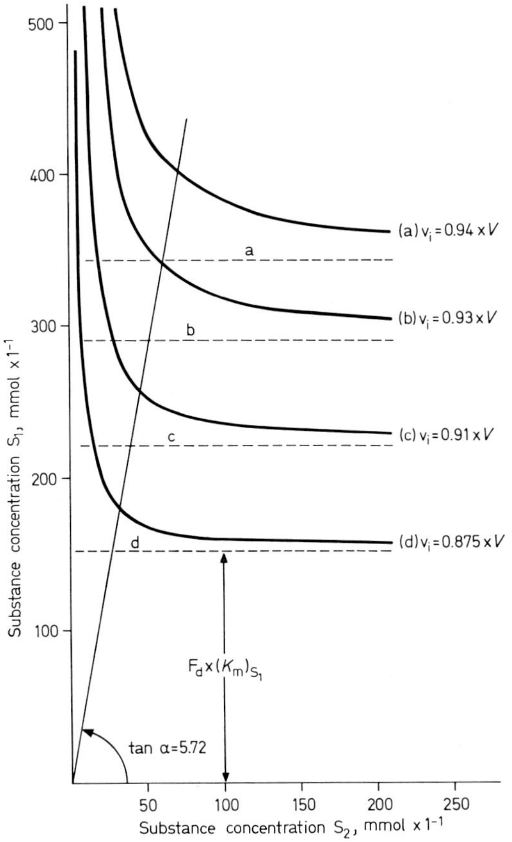

Fig. 2. Two-substrate reaction alanine aminotransferase. Interdependency of substrate concentrations. S_1 = L-alanine; S_2 = 2-oxoglutarate. $(K_m)S_1 = 21.9$ mmol/*l*; $(K_m)S_2 = 0.67$ mmol/*l*. Values of $F = v_i/V - v_i$ are: (a) 15.67; (b) 13.29; (c) 10.11; (d) 7.00. $\tan \alpha = 5.72$.

if $[I] = [S_2]$ it follows:

$$[S_1] = \frac{(K_m)_{S_1} \times \left(1 + \frac{[S_2]}{K_I}\right)}{\frac{1}{F} - \frac{(K_m)_{S_2}}{[S_2]}}.$$

This is a quadratic relationship between $[S_1]$ and $[S_2]$. Therefore, it can be demonstrated by a parabola. In figure 3, it is shown how the hyperbola for $v_i = 0.91 \times V$ of the ALT reaction is transformed to a parabola when the inhibitor term is included.

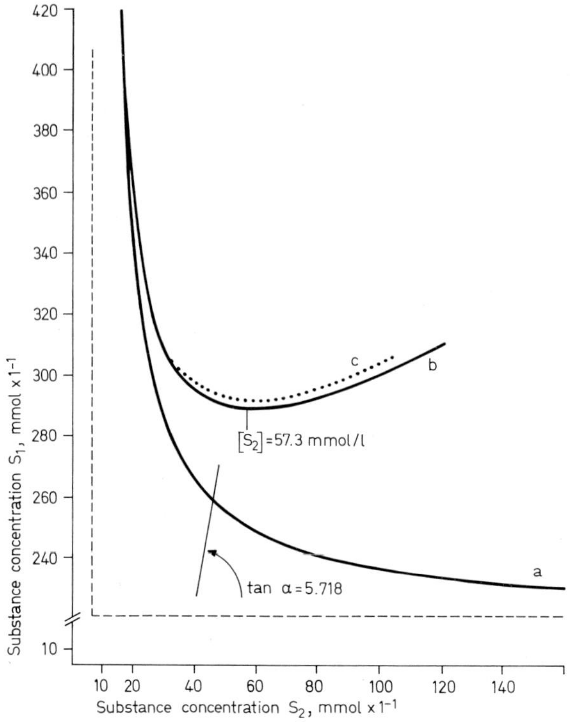

Fig. 3. Alanine aminotransferase. Interdependency of substrate concentrations. S_1=L-alanine; S_2=2-oxoglutarate. $(K_m)_{S_1}=21.9$ mmol/*l*; $(K_m)_{S_2}=0.67$ mmol/*l*. F=10.11 corresponding to $v_i=0.91 \times V$. *a* Calculated with the above data, inhibition not considered. *b* Calculated with the above data, competitive inhibition by 2-oxoglutarate considered ($K_I=400$ mmol/*l*). *c* Calculated with the above data according to *b*, additionally product inhibition by L-glutamate considered ($K_I=14$ mmol/*l*).

Calculation of the optimum substrate concentration pair of a two-substrate enzyme reaction on the basis of absolute K_m values is a better way than the empirical plot of v_i against increasing substrate concentrations. This, then, is the third point we learnt.

Indicator Reactions

The indicator enzyme also reacts with two substrates. One is NADH, called S_3, the other is, in the case of AST, oxaloacetate, called P_1. Primary and indicator reaction must reach a steady state soon in which:

$$0.96 \times V_{\text{prim.}} = (v_{\text{eff.}})_{\text{ind.}}$$

In the case of AST, the fraction v_i over V has been chosen to be 0.96; a higher value is no longer convenient. For the effective reaction rate of the indicator reaction, we once more can apply our rate law for two-substrate enzyme reactions in which, instead of $[S_1]$, now $[P_1]$ stands and in which V represents the number of units of malate dehydrogenase.

It can easily be shown by a rough calculation that several simplifications can be made. Under measurement conditions, the indicator reaction follows zero order kinetics and we can calculate how many units of indicator enzyme relative to the units of the primary enzyme must be used in the assay system.
Finally we get:

$$V_{ind.} = (1 + (K_m)_{P_1}/[P_1]) \times 0.96 \times V_{prim.}$$

For numerical values one must know $[P_1]$. This is nothing else than the steady-state concentration at the end of the lag phase of this reaction. It can be fixed arbitrarily within certain limits. If, for example, we tolerate only a small lag phase up to OD=0.006, this corresponds to a steady-state concentration of 0.001 mmol/*l*. We have to check if reasonable conditions for the indicator reaction can be realised with this value:

If the K_m of malate dehydrogenase with respect to oxaloacetate is:

$$(K_m)_{P_1} = 0.0165 \text{ mmol}/l,$$

and if the steady-state concentration of oxaloacetate is limited to:

$$[P_1] = 0.001 \text{ mmol}/l,$$

then the ratio of both is:

$$(K_m)_{P_1}/[P_1] = 16.5,$$

and the maximum velocity of the indicator reaction is given by:

$$V_{ind} = 17.5 \times 0.96 \times V_{prim.} = 16.8 \times V_{prim.}$$

If we want to have a method valid for serum up to 300 U/*l*, we have, for example, 25 U/*l* in the assay mixture. The catalytic activity concentration of the indicator enzyme then must be 16.8 times greater, that means 420 U/*l*. That is convenient.

As the fourth point, we have learnt that in a coupled reaction sequence the constant steady-state concentration of the primary reaction product to be measured in the indicator reaction is a measure of the lag phase. It can be fixed arbitrarily within certain limits.

The smaller the lag phase shall be, the smaller $[P_1]$ must be. With decreasing $[P_1]$, the ratio $(K_m)_{P_1}/[P_1]$ rises. That means nothing else than the well-known fact that with increasing catalytic concentration of the indicator enzyme the lag phase decreases.

In this way, the basic design of two coupled two-substrate enzyme reactions can be made clear. The advances in the theoretical treatment of enzyme reactions in the last few years have contributed a great deal to the developments in methodology of enzyme assays in clinical chemistry. It is to be hoped that further evaluation of the design of new methods or of revision of existing methods in order to specify reference methods will follow this or similar approaches.

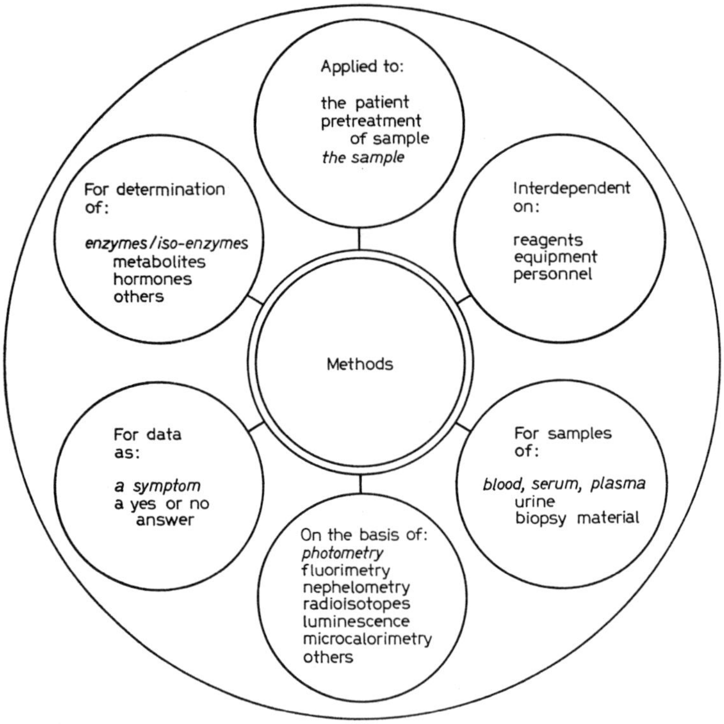

Fig. 4. Survey about methods in clinical chemistry.

Conclusion

From the huge variety of aspects of methodology in clinical chemistry, I could only select a few (fig. 4). So I have tried to show here methods for the photometric determination of enzymes or iso-enzymes in samples of blood, serum or plasma in order to provide data which can serve in clinical diagnosis.

References

1 BERGMEYER, H. U.: Neue Werte für die molaren Extinktions-Koeffizienten von NADH und NADPH zum Gebrauch im Routine-Laboratorium. Z. klin. Chem. klin. Biochem. *13:* 507–508 (1975).

2 BERGMEYER, H. U.; SCHEIBE, P., and WAHLEFELD, A. W.: Optimization of methods for aspartate aminotransferase and alanine aminotransferase. Clin. Chem. *24:* 58–73 (1978).

3 BERGMEYER, H. U.: Evaluation of optimum conditions of two-substrate enzyme reactions. J. clin. Chem. clin. Biochem. *15:* 405–410 (1977).

4 BERGMEYER, H. U.: in BERGMEYER Grundlagen der enzymatischen Analyse, pp. 58–81 (Verlag Chemie, Weinheim 1977).

5 BUETTNER, H.: Geschichte der Klinischen Chemie. Med. Welt. *28:* 1238–1243 (1977).

6 DOSS, M. und TIEPERMANN, R. VON: Enzymopathie der Porphyrinsynthese in Erythrozyten bei akuter hepatischer Porphyrie. Lab. Med. *2:* 28–32 (1977).

7 Expert Panel on Enzymes; Committee on Standards (IFCC): Provisional Recommendations on IFCC methods for the measurement of catalytic concentrations of enzymes. 2. IFCC method for aspartate aminotransferase. Clin. chim. Acta *70:* p. 336 ff., F19–F42 (1976).

8 JAMES, G. P.; PASSEY, R. B.; FULLER, J. B., and GILES, M. L.: Evaluation of the Du Pont aca α-amylase procedure. Clin. Chem. *23:* 546–550 (1977).

9a LAUDAHN, G.; HARTMANN, E.; ROSEFELD, E. M.; WEYER, H., and MUTH, H. W.: Normalwerte der Serum-Transaminasen bei Verwendung substratoptimierter Testansätze zur Aktivitätsmessung. Klin. Wschr. *48:* 838 (1970).

9b LONDON, J. W.; YARRISH, R.; DZUBOW, L. D., and GARFINKEL, D.: Computer simulation and optimization, as examplified by the enzyme-coupled aminotransferase (transaminase) assays. Clin. Chem. *20:* 1403–1407 (1974).

9c LONDON, J. W.; SHAW, L. M.; FETTEROLF, D., and GARFINKEL, D.: A systematic approach to enzyme assay optimization, illustrated by aminotransferase assays. Clin. Chem. *21:* 1939–1952 (1975).

10 NEUMEIER, D.; PRELLWITZ, W.; WÜRZBURG, U.; BRUNDOBLER, M.; OLBERMANN, M.; JUST, H. J.; KNEDEL, M., and LANG, H.: Determination of creatine kinase isoenzyme activity in serum using immunological inhibition of creatine kinase M subunit activity. Activity kinetics and diagnostic significance in myocardial infarction. Clin. chim. Acta *73:* 445–451 (1976).

11a PERSIJN, J. P. and SLIK, W. VAN DER: A new method for the determination of γ-glutamyltransferase in serum. J. clin. Chem. clin. Biochem. *14:* 421–427 (1976).

11b THEODORSEN, L. and STRØMME, J.H.: γ-Glutamyl-3-carboxy-4-nitroanilide: the substrate of choice for routine determinations of γ-glutamyltransferase activity in serum? Clin. chim. Acta *72:* 205–210 (1976).

12a SZASZ, G.; GRUBER, W., and BERNT, E.: Creatine kinase in serum. 1. Determination of optimum reaction conditions. Clin. Chem. *22:* 650–656 (1976).

12b SZASZ, G.; GERHARDT, W.; GRUBER, W., and BERNT, E.: Creatine kinase in serum. 2. Interference of adenylate kinase with the assay. Clin. Chem. *22:* 1806–1811 (1976).

13 THEFELD, W.; HOFFMEISTER, M.; BUSCH, E.W.; KOLLER, P.U. und VOLLMAR, J.: Referenzwerte für die Bestimmungen der Transaminasen GOT und GPT, sowie der alkalischen Phosphatase im Serum mit optimierten Standartmethoden. Dt. med. Wschr. *99:* 343–351 (1974).

14 WITT, I.: Neue Methoden der Gerinnungsanalyse mit chromogenen Substanzen. J. clin. Chem. clin. Biochem. *15:* 239–244 (1977).

15 ZIEGENHORN, J.; SENN, M., and BUECHER, T.: Molar absorptivitites of β-NADH and β-NADPH. Clin. Chem. *22:* 151–160 (1976).

H. U. BERGMEYER, PhD, Boehringer Mannheim Research Center, *D-8132 Tutzing* (FRG)

Enzymes in Health and Disease. Inaug. Scient. Meet. Int. Soc. Clin. Enzymol., London 1977, pp. 87–92 (Karger, Basel 1978)

Standardization of Enzyme Assays: the Case of Amylase

A. BURLINA and L. GALZIGNA

Center for Clinical Enzymology, Department of Clinical Chemistry, Centro Ospedaliero Borgo Trento, Verona, and Institute of Biological Chemistry, University of Padova, Padova

The details of the general procedures for standardization of enzyme assays have been issued by IFCC [1975] and several provisional recommendations for single enzyme assays have appeared so far.

We shall not review here what can be found in the literature, including the problem of the temperature of assay which is one of the most debated topics in enzyme assay standardization. In most of the Proceedings of Meetings of Clinical Enzymology, detailed discussions on the problem of temperature and other problems of standardization can be found [COPELAND *et al.*, 1977].

α-Amylase (1,4-α-D-glucanglucanohydrolase, EC 3.2.1.1) determination in serum has been chosen as the theme of the present paper because we are now establishing a new procedure for a continuous assay of this enzyme. Serum amylase is very often an emergency determination and a suitable procedure with satisfactory characteristics from the standpoint of enzyme assay standardization is still awaited. In fact, no reference method has been established and difficulties are created by the choice of a true physiological substrate and by the possibility that other enzymes, such as maltase, interfere with the determination of what we consider serum amylase.

Methods developed hitherto for the assay of amylase include the so-called amyloclastic procedure with starch-iodide [SOMOGYI, 1960], dyed-starch or chromolytic methods [KLEIN *et al.*, 1970], saccharogenic methods [HENRY and CHIAMORI, 1960] for the measurement of reducing groups liberated by amylase action, coupled enzymatic assays [KESTON and BRANDT, 1963] and methods based on light-scattering techniques [LEHANE *et al.*, 1977]. The main problems for accepting any of the above procedures as a reference method include variability of the starch substrate used, complex and sometimes time-consuming manipulations, interferences with serum components, stability of the reagents, etc.

None of the proposed procedures is free from inconvenience and the main disadvantage of all except the coupled enzymatic assay is the fact that they are multi-point and not continuous assays. We do not yet have sufficient experience of coupled enzymatic assays and in this paper coupled assays and turbidimetric assays will not be considered.

Another important problem, common to all procedures, which introduces additional difficulties in the assessment of amylase results is that the distribution of values relative to reference and patient populations is not gaussian. This makes it necessary to use non-parametric statistical analysis of the values for both serum and urine and to elaborate the data according to the Kolmogorov-Smirnov or similar procedures.

We have carried out parallel checks of two chromolytic methods with a slightly modified version of the amyloclastic procedure, carried out with 5 min incubation and pure potato amylose as substrate.

The results constitute preliminary data for a large survey which will be completed over a long period of time.

The amyloclastic procedure was carried out according to SOMOGYI [1960] and the absorbance change at 578 nm was transformed into $\Delta A/\text{min}$ for the calculation of international units (IU). The extinction coefficient of the amylose-iodine solution was calculated after measuring its absorption spectrum. The molecular weight of the potato amylose (type II, Sigma) was measured by a viscosimetric method according to EVERETT and FOSTER [1959] and a mean value of 115000 was obtained. In this way, a molar extinction coefficient of 1×10^4 was obtained for iodine, whereas the extinction coefficient of the amylose-iodine complex was 4.34×10^5. The factor used to transform the absorbance values into IU can be therefore either 32500 (if molar extinction coefficient of iodine is used) or 750 (if molar extinction coefficient of the amylose-iodine complex is used).

The yield of the amyloclastic method was compared with that of the saccharogenic method carried out according to HENRY and CHIAMORI [1960] using pure hog α-amylase (Worthington) and the amyloclastic method was found to have lower sensitivity.

The comparison between two chromolytic methods (i.e. the commercial kits of the so-called Amylochrome, Roche, and Dyamyl-L, General Diagnostics) and the amyloclastic method was then carried out. Two groups of adult male patients, i.e. a group of orthopaedic and a group of pancreatic patients were considered. Serum amylase was measured by Amyloclastic, Amylo-

Table I. Serum amylase and its correlation with total protein

	Control group (n=10)	r′	Pancreatic patients (n=10)	r′
Amylochrome, IU/l	219±60.6	0.92	699.8±459	1.00
Dyamyl-L, DU	113±40.6	1.00	453.5±336	1.00
Amyloclastic, IU/l	180.7±68.3	1.00	831.3±672	0.90
	4.1±1.5		18.9±15.3	

Amylase values are given as mean±SD in either IU or Dyamyl units. The rank correlation (r′) is calculated by means of the Spearman's test as $r' = 1 - (6\Sigma d^2/n[n^2-1])$.
Two values are given for the amyloclastic method depending on the two different factors of conversion used.

Table II. Normal ranges, precision of the methods and apparent amylase activity of albumin

	Normal range	CV% (within run)	Apparent activity of albumin
Amylochrome, IU/l	140–310	2.19	30
Dyamyl-L, DU	67–198	3.90	18
Amyloclastic, IU/l	117–325	5.6	65
	2.65–7.41	5.6	1.47

The sample of bovine serum albumin has a volume equal to that of serum and its apparent amylase activity is measured with the three methods.

chrome and Dyamyl-L methods and total serum protein content was determined according to GORNALL *et al.* [1949]. The rank correlation test of Spearman for small samples was applied to evaluate the correlation between protein content and amylase values.

A significant correlation between total protein content and serum amylase values determined with all the methods was found (table I).

Table II shows the normal ranges observed and the variation coefficients relative to the three methods. The apparent amylase activity of a sample containing 7% of bovine serum albumin (fraction V, Miles) was also measured and with all three procedures significant amylase activity was found due to albumin alone.

Table III. Dilution of serum samples with addition of pure hog amylase

	Undiluted	1:2	1:4	1:16
1st sample	1,050	455	227.5	122.5
	23.8	10.3	5.1	2.7
2nd sample	997.5	490	280	87.5
	22.6	11.1	6.3	1.9
3rd sample	1,032.5	490	262.5	70
	23.4	11.1	5.6	1.5

15.56 IU of amylase was added to the samples after inactivation. The amyloclastic method was used to measure amylase activity.

Table IV. Effect of temperature on amylase activity of four commercial control sera

	Amylochrome, IU/l			Dyamyl-L, Dyamyl units			Amyloclastic, IU/1		
	37°C	30°C	25°C	37°C	30°C	25°C	37°C	30°C	25°C
Ortho-normal	208	140	110	110	75	35	228	162	65
							5.25	3.75	1.5
Ortho-abnormal	254	240	159	185	158	97.5	325	228	162
							7.5	5.25	3.75
Versatol-E-N-Plus	164	139	101	204	162	100	195	130	78
							4.5	3.0	1.8
Versatol-E-Plus	362	238	162	465	306	202	292	195	130
							6.75	4.5	3.0

Table III shows the activity found after dilution of three samples of serum inactivated 20 min at 56 °C to which the same amount of purified hog amylase was added (0.166 μl corresponding to 4.095 IU measured by the saccharogenic method and 15.56 IU measured by the amyloclastic method).

The above data demonstrate that the use of the extinction coefficient of the amylose-iodine complex allows us to obtain the results, expressed as IU, that are closer to theoretical values. The activity found with the amyloclastic method (table III) is higher than the theoretical value because the effect of serum albumin is present. The values of amylase activity calculated by use of the extinction coefficient of amylose-iodine are of the order of magnitude of those obtained with the coupled enzymatic assay [Smith-Kline, 1977].

The effect of different temperatures on the amylase activity of four control sera measured with the three methods is shown in table IV.

Among the methods used by us, the amyloclastic procedure seems the most sensitive to the temperature change and the Amylochrome the least sensitive, as can be judged from the ratios of the values obtained at the different temperatures.

The different methods used in this work seem to have the same disadvantages, the most important of which appears to be the effect of serum protein. Our version of the amyloclastic procedure is suitable for emergency determinations, but obviously does not eliminate the inconveniences that have been fully discussed by other authors [SEARCY *et al.*, 1965]. The three methods, on the other hand, seem to assay true amylase activity, i.e. the activity involved in the splitting of large molecules.

Our efforts towards the standardization of amylase assay should begin therefore from the optimization of methods of this type for which the obvious reference method is the amyloclastic method.

References

COPELAND, B.E.; DOHERTY, A.; JORDAN, J., and PARKEY, R.: Evaluation of a new turbidimeter for serum amylase measurements using a turbidity reference material and a stable pectin substrate; in BURLINA and GALZIGNA Clinical enzymology symposia, vol. 1, pp. 183–188 (Piccin, Padua 1977).

EVERETT, W.W. and FOSTER, J.F.: The conformation of amylose in solution. J.Am. chem.Soc. *81:* 3464–3469 (1959).

GORNALL, A.G.; BARDAWILL, C.J., and DAVID, M.M.: Determination of serum protein by means of biuret reaction. J.biol.Chem. *177:* 751–766 (1949).

HENRY, R.J. and CHIAMORI, W.: Study of the saccharogenic method for the determination of serum and urine amylase. Clin.Chem. *6:* 434–452 (1960).

IFCC: Methods for the measurement of catalytic concentration of enzymes. Clin.chim. Acta *61:* 238 ff. (1975).

KLEIN, B.; FOREMAN, J.H., and SEARCY, R.L.: New chromogenic substrate for determination of serum amylase activity. Clin.Chem. *16:* 32–38 (1970).

KESTON, A.S. and BRANDT, R.: Methods of analysis of the anomers of glucose. Analyt. Biochem. *6:* 461–467 (1963).

LEHANE, D.P.; WISSERT, P.J.; LUM, G., and LEVY, A.L.: Amylase activity in serum and urine. Comparison of results by the amyloclastic, dyed-starch and nephelometric techniques. Clin.Chem. *23:* 1061–1065 (1977).

SEARCY, R.L.; HAYASHI, S.; HARDY, E.M., and BERK, J.E.: The interaction of human serum protein fractions with the starch-iodine complex. Clin. chim. Acta *12:* 631–638 (1965).

Smith-Kline Instruments: Tech. Bull. on Eskalab Amylase Bulk Reagent (1977).

SOMOGYI, M.: Modifications of two methods for the assay of amylase. Clin. Chem. *6:* 23–29 (1960).

A. BURLINA, MD, Centro Ospedaliero Borgo Trento, Center for Clinical Enzymology, Department of Clinical Chemistry, *I-37100 Verona* (Italy)

Enzymes in Health and Disease. Inaug. Scient. Meet. Int. Soc. Clin. Enzymol., London 1977, pp. 93–97 (Karger, Basel 1978)

The Multivariate Approach Applied to Enzyme Levels Interpretation

R. J. WIEME, J. VAN EGMOND and G. DE SCHRIJVER

Laboratory for Clinical Pathology, University of Ghent, Ghent

Clinical chemists are in a position inferior to that of clinicians when it comes to *real* interpretation of results. However, we think that in the near future a profound change will occur in this situation and that a new stimulus will be found for motivating young medical doctors to engage in clinical chemistry, thus correcting a basic shortcoming stressed by ASTRUP [1]. Indeed, the complete medical record is progressively developing into an on-line database, available all over the hospital to all health-care workers that do have access. The second reason is that we are providers of many quantitative data that cannot be exploited in any other way than by multivariate analysis. The range of normal variation when two uncorrelated tests are considered together is larger than, and has a shape different from, the square obtained when the two test results are represented on a normalized base. If the two results present correlation the situation is changed even more dramatically and most combinations that, on a univariate base, should be considered as normal, become improbable.

Another way of introducing the concept of multivariate analysis is by referring to multiphasic health screening. The probability of getting at least one 'abnormal' test result is quite high if five tests are requested ($p \approx 0.5$). This has led some investigators to the conclusion that they should increase the reference range to 3 σ to reduce false alarms. In point of fact, they should have analyzed the results against a pentavariate normal distribution [12].

In a trivariate system, the position of a patient (the patient vector $\vec{P}$) is defined by three tests. For a n-variate system the $\vec{P}$ is positioned in a n-dimensional Euclidian space and describes the patient according to the investigated parameters. A clustering set of $\vec{P}$s can be considered an operational description of a disease ($\vec{D}$: disease vector). $\vec{D}$ is characterized by a centroid parameter and by a measure of the distribution.

In diagnostic enzymology, the bivariate approach has already found a widespread application by the use of the De Ritis ratio (AST/ALT). But several other ratios (e.g. CK/AST, γ-GT/AST) have been proposed. Enzyme ratios have also been made more complex by increasing the number of tests included and also by further differentiating the scale of values (ratio AST + ALT to GLDH).

However, for several reasons, the practical value of most enzyme ratios is limited. (1) One does not consider levels. (2) Precision decreases and the problem of accuracy becomes acute (different ratio values in different laboratories). (3) Considerations valid for elevated levels do not necessarily apply to the normal values. A value of 1.2 is found for the De Ritis ratio at normal levels. (4) Difficulties occur in making measurement at low and also at high levels. (5) Differential stability of enzymes upon storage can change the ratio. For most enzymes, no changes are expected to occur when the serum is kept at 4 °C. (6) The ratio may change in the course of the disease. This stresses the need for considering the stage of the disease. If not taken into account, this may constitute a very important source of 'noise' when one is searching for 'clusters' in a distribution of $\vec{P}$s.

What else can we expect from elaborated multivariate analysis?

(1) Among the signs, symptoms and test results observed in patients, it can detect structure (clustering). The significance remains to be defined ('pattern cognition') [13].

(2) Already in possession of knowledge of disease patterns, it can adjust the choice and weight of signs, symptoms and of test results so as to cluster the patients in accordance with medical knowledge ('pattern recognition').

(3) Multivariate analysis can develop procedures of algorithms for classifying, using maximal economy and with known probability. This procedure can be called 'pattern identification' and is actually congruent with diagnosis making.

In all multivariate analyses, there comes a moment where one has to judge the quality of the cluster, or the position of $\vec{P}$ relative to $\vec{D}$s. This is obtained by a measure of 'distance' and by a measure of spreading.

For the former, Euclidian distance between vector points or, better still, the square of that distance is used. For the latter, variance is generally the measure of spreading and can be considered to be composed of an intra-cluster and of a between-cluster component. Obviously, optimal resolution is obtained when the separation is such that the inter-cluster variance is rendered as large as possible. A major difficulty arises for the non-gaussian distributions that are the rule and non-parametric tests are just beginning to be applied.

For daily practice, a 'discriminant function' is derived for the system under study. Most of the time this is a linear equation, with or without weighting of the parameters. According to its resultant value, a diagnosis is accepted or rejected, but one has to take into account *a priori* probabilities, if the probability of the disease for the population under consideration is low (say $p<0.05$).

A typical example is seen in the discriminant function for distinguishing 'normal' from 'liver disease', using 4 simultaneous tests and this with $p=0.85$ for success [7]. Each of the selected tests alone shows marked overlap between the normal and pathological ranges.

This recent study is strikingly similar to the pioneer work of ZIEVE and HILL [14] in 1955. These authors tried to separate patients with liver cirrhosis from normals. Only four of the eleven tests used (BSP, $ZnSO_4$, hippuric acid and urine coproporphyrin) remained of interest. CHARBONNIER *et al.* [4] used discriminant analysis of serum proteins for the differentiation of icteric states, and others studied combinations of enzyme tests with other laboratory parameters [2, 3, 6, 9–11].

In such investigations, it is difficult to collect a series of patients who have been submitted to all tests under study. All authors agree that, of the numerous tests examined, only a few prove of real discriminatory value [8].

However, the points seem to us to have been insufficiently explored. Redundancy in tests can have other justifications: (1) acquisition of convergent information can protect against laboratory error; (2) closely correlating tests could not be expected to be of enhanced value when combined, but non-correlation might have special significance.

The tests that proved to be the most useful are alkaline phosphatase (AP), followed by AST and ALT. Then come specific protein assays and functional tests of the type bromsulphalein and galactose elimination. Isozymes have been little investigated, except for the LDH system [5]. It could be interesting to use also the AP and CK systems.

Summary

In view of the increasing importance of clinical chemistry in diagnosis, further investigation of multivariate analysis, or other taxonomic procedures, is urgent. As yet, there seems to be little reward in its use, except perhaps in high-risk dichotomous situations, such as the decision between surgical and medical icterus. The main impediments to its development that we see are the difficulties associated with quantitative interpretations of probability when distributions are non-gaussian, as is generally the case.

Computers are rapidly becoming more powerful, and we now know at least some of the pitfalls that await us. Investigations should be prospective rather than retrospective and they should take into account the fact that medical reality corresponds to a multitude of low frequency possibilities where a decision with 5% uncertainty is quite inadequate.

References

1 ASTRUP, P.: Clinical chemistry as a medical discipline. Scand. J. clin. Lab. Invest. *37:* 1–5 (1977).
2 BARON, D. and FRASER, P.: Medical applications of taxonomic methods. Br. med. Bull. *24:* 236–240 (1968).
3 BURBANK, F.: A computer diagnostic system for the diagnosis of prolonged undifferentiating liver disease. Am. J. Med. *46:* 401–415 (1969).
4 CHARBONNIER, A.; CYFFERS, B.; SCHWARTZ, D. et VESSERAC, A.: Discrimination entre ictères médicaux et chirurgicaux à partir des résultats de l'analyse électrophorétique des protéines du sérum. Revue int. Hépat. *7:* 1039–1064 (1955).
5 GLICK, J.: Serum lactate dehydrogenase isoenzyme and total lactate dehydrogenase values in health and disease, and clinical evaluation of these tests by means of discriminant analysis. Am. J. clin. Path. *52:* 320–328 (1969).
6 GRAMS, R.; JOHNSON, E., and BENSON, S.: Laboratory data analysis system. II. Multivariate diagnosis. Am. J. clin. Path. *58:* 201–207 (1972).
7 HAMILTON, M.: A simple discriminant function for hepatic disease. J. clin. Path. *30:* 454–459 (1977).
8 PLOMTEUX, G. et TOULET, J.: Sélections des variables biochimiques discriminantes pour le diagnostic des ictères par cholostase intra-hépatique, par obstacle extra-hépatique ou par cancer du foie. Archs fr. mal. app. Dig. *64:* 317–324 (1975).
9 RAMSOE, K.; TYGSTRUP, N., and WINKEL, P.: The redundancy of liver tests in the diagnosis of cirrhosis estimated by multivariate statistics. Scand. J. clin. Lab. Invest. *26:* 307–312 (1970).
10 SOLBERG, H.; SKREDE, S.; ELGIO, K.; BLOMHOFF, J.P., and GIONE, E.: Classification of liver diseases by clinical chemical laboratory results and cluster analysis. Scand. J. clin. Lab. Invest. *36:* 81–85 (1976).
11 TENGSTRÖM, B.: The discriminatory ability of a galactose tolerance test and some other tests in the diagnosis of cirrhosis of the liver, hepatitis and biliary obstruction. Scand. J. clin. Lab. Invest. *23:* 159–168 (1969).
12 WINKEL, P.; LYNGBYE, J., and JÖRGENSEN, K.: The normal region – a multivariate problem. Scand. J. clin. Lab. Invest. *30:* 339–344 (1972).

13 Winkel, P.: Patterns and clusters – multivariate approach for interpreting clinical chemistry results. Clin. Chem. *19:* 1329–1338 (1973).
14 Zieve, L. and Hill, E.: An evaluation of factors influencing the discriminative effectiveness of a group of liver function tests. III. Relative effectiveness of hepatic tests in cirrhosis. Gastroenterology *28:* 785–801 (1955).

R. J. Wieme, MD, University of Ghent, Academic Hospital, Department of Clinical Pathology, De Pintelaan 135, *B-9000 Ghent* (Belgium)

Enzymes in Health and Disease. Inaug. Scient. Meet. Int. Soc. Clin. Enzymol., London 1977, pp. 98–102 (Karger, Basel 1978)

Standardization of Enzyme Activities Determination in France

M. Mathieu, M. Bailly, J. P. Bretaudiere, C. Dorche, M. M. Galteau, J. Guidollet, P. Laleigerie, P. Louisot, J. C. Leunis, G. Plomteux, F. Schiele and M. Vancraeynest

Enzyme Commission of the 'Société Française de Biologie Clinique'

The Enzyme Commission (EC) of the Société Française de Biologie Clinique has been working for 4 years to promote the standardization of enzyme activity determination in clinical chemistry.

Up to now, four documents have been published concerning the choice of temperature and selected methods for aspartate aminotransferase, creatine kinase and alkaline phosphatase. A proposed method for alanine aminotransferase is now being elaborated. In this paper, we describe our working scheme, the results obtained and our future aim.

I. Working Scheme

Table I describes the organization of French Society of Clinical Biology. Under the administration and direction boards, the standard committee controls ten specialized commissions: the chairman of each commission is a member of the Standard Committee (SC) and is often also an 'associate member' of the Expert Panel of the International Federation of Clinical Chemistry. Usually, one commission includes 5–10 members.

In addition to the specialized commission, the Standard Committee has created 'Working Groups' to deal with special problems.

In table II, we describe the working scheme of the EC: worklines are selected on theoretical and practical grounds by the members of the commission, then submitted to the SC; after acceptance, experimental studies are performed by the EC members.

Results are then discussed and controlled in the different laboratories

Table I. Société Française de Biologie Clinique

1. Administration Board
↓
2. Direction Board
↓
3. Standard Committee
↓
4. Specialized Commissions: teaching, reference values, instrumentation, proteins, immunoglobulins, hormones, quality control, automation, *enzymes*, lipids-lipoproteins, drugs
5. Working Groups: glucose, serum iron, phosphate, uric acid

Table II. Société Française de Biologie Clinique: Enzyme Commission

1. Documents: stage 1
2. Approval by Standard Committee: stage 2
3. Diffusion for criticism to:
French biologists by *L'Information Scientifique du Biologiste*
Working groups of other countries
Apparatus and reagents manufacturers
4. Final approval by Standard Committee
5. Recommendations (stage 3) published in *Annales de Biologie Clinique*

of the group and the 'stage I' documents are elaborated. These documents are submitted to the SC for approval and then constitute stage 2. Document 2 is distributed for criticism to French Biologists by the means of *L'Information Scientifique du Biologiste*, to working groups in other countries, and to apparatus and reagent manufacturers. After receiving suggestions and criticisms, document 3 is prepared and submitted for final approval to the SC.

The final recommendations are published in *Les Annales de Biologie Clinique*.

II. Recommended Methods in 1977

Four documents [1–4] were published during the last 3 years and are presented at stage 3 as 'recommendations': First [1], we proposed 30 °C as standard temperature used for the measurement of usual enzyme activity in

Table III. Aspartate aminotransferase (EC 2.6.1.1)

	SFBC 1976 [2, 3]	IFCC 1975–1977 [6]
Temperature, °C	30	
pH (at 30 °C)	7.8	7.8
Tris (hydroxymethyl) aminomethane	80 mmol/l^{-1}	80 mmol/l
L-Aspartate	200 mmol/l^{-1}	240 mmol/l
2-Oxoglutarate, mmol/l^{-1}	12	12
NADH, mmol/l^{-1}	0.18	0.18
Pyridoxal phosphate, mmol/l^{-1}	0.10	0.10
Malate dehydrogenase	600 μmol/min^{-1}/l^{-1}	10 μmol/sec^{-1}/l^{-1}
Lactate dehydrogenase	900 μmol/min^{-1}/l^{-1}	10 μmol/sec^{-1}/l^{-1}
Volume fraction of sample	1/12	1/12

Table IV. Creatine kinase (EC 2.7.3.2)

	SFBC 1976 [2, 3]	Scandinavian 1976 [8]	German 1977 [7]
Temperature, °C	30	37	25
pH	6.6	6.5	6.7
Imidazole acetate, mmol/l^{-1}	100	100	100
Creatine phosphate, mmol/l^{-1}	30	30	30
N-Acetylcysteine, mmol/l^{-1}	20	20	20
ADP, mmol/l^{-1}	2	2	2
D-Glucose, mmol/l^{-1}	2	2	2
Magnesium acetate, mmol/l^{-1}	10	10	10
NADP, mmol/l^{-1}	2	2	2
AMP, mmol/l^{-1}	5	5	5
Diadenosine pentaphosphate, μmol/l	0	10	10
Hexokinase, U/l	3,000	3,500	2,500
D-Glucose-6-phosphate dehydrogenase, U/l	2,000	2,000	1,500
Volume fraction of sample	1/30	0.043	0.04

Table V. Alkaline phosphatase (EC 3.1.3.1). SFBC, 1977 [4]

Temperature, °C	30
pH (at 30 °C)	10.5 ± 0.05
2-Amino-2-methylpropanol, mmol/l^{-1}	900
4-Nitrophenylphosphate (disodium salt), mmol/l^{-1}	16
Magnesium sulphate, 7 H_2O, mmol/l^{-1}	1
Volume fraction of sample	1/30

Table VI. Alanine aminotransferase (EC 2.6.1.2). SFBC, 1977 [5]

Temperature, °C	30
pH (at 30 °C)	7.4±0.1
Tris (hydroxymethyl) aminomethane, mmol/l^{-1}	80
L-Alanine, mmol/l^{-1}	400
2-Oxoglutarate, mmol/l^{-1}	18
Pyridoxal phosphate, mmol/l^{-1}	0.1
NADH, mmol/l^{-1}	0.18
Lactate dehydrogenase, mmol × min^{-1} × l^{-1}	2
Volume fraction of sample	1/12

clinical chemistry. The following documents [2–4] concern our recommendations for the following catalytic activities: aspartate aminotransferase (table III), creatine kinase (table IV) and alkaline phosphatase (table V).

An additional document [5] at stage 2 is being distributed for criticism as a proposal for the measurement of alanine aminotransferase (table VI) activity. A further document is now at stage 1 concerning a proposal for a recommended method for measurement of catalytic activity of lactate dehydrogenase. Tables IV–VI illustrate the three recommended methods, compared with those of other countries.

III. Aims of the Enzyme Commission

We are now working at the elaboration of the second document concerning each of the recommended methods. This second document deals with the applicability of those methods, that is to say: reference values, the way of keeping samples, reagent purity, influencing factors such as diluting and anticoagulants. The next enzyme catalytic activities which will be studied are- γ-glutamyl transpeptidase, amylase and 5′- nucleotidase.

Much work remains, but it is very difficult to progress rapidly as many resources are necessary: research, apparatus, reagents.

Furthermore, we wish that our recommendations might be in agreement with other countries and finally with the International Federation of Clinical Chemistry.

The constant evolution of research, especially in the field of 'optimization' in enzyme catalytic activity measurement makes it very difficult to propose a standardized method at a precise time.

But the first thing we have to keep in mind is the urgent need for standardization in this field, and this must be proposed quickly. One must find methods, relatively simple to use by all biologists, inexpensive, but reliable and reproducible.

Summary

The Enzyme Commission of the 'Société Française de Biologie Clinique' has been working for 4 years to promote the standardization of enzyme activity determination in France. Up to 1977, four documents have been published concerning recommended methods (aspartate aminotransferase, EC 2.6.1.1, creatine kinase EC 2.7.3.2, alkaline phosphatase EC 3.1.3.1) and choice of temperature (30 °C). A further method has now been proposed (alanine aminotransferase EC 2.6.1.2).

References

1 Société Française de Biologie Clinique: Recommandations pour la standardisation de la température utilisée pour la mesure des activités enzymatiques courantes en chimie clinique. Annls Biol. clin. *33:* 51 (1975).
2 Recommandations pour la mesure de l'activité catalytique de l'aspartate aminotransférase dans le sérum à 30 °C. Annls Biol. clin. *34:* 291 (1976).
3 Recommandations pour la mesure de l'activité catalytique de la créatine kinase dans le sérum humain à 30 °C. Annls Biol. clin. *34:* 297 (1976).
4 Recommandations pour la mesure de l'activité catalytique des phosphatases alcalines dans le sérum humain à 30 °C. Annls Biol. clin. *35:* 271 (1977).
5 Proposition de recommandation pour la détermination dans le sérum humain de la concentration catalytique de l'alanine aminotransférase à 30 °C. Inform. Sci. Biol. *3:* 22 (1977).
6 International Federation of Clinical Chemistry: Committee on Standards. Expert Panel on Enzymes Clin. chim. Acta *61:* F 11 (1975); Clin. chim. Acta *70:* F 19 (1976); Clin. Chem. *23:* 887 (1977).
7 Recommendations of the German Society for Clinical Chemistry: Standard method for the determination of creatine kinase activity. J. clin. Chem. clin. Biochem. *15:* 255 (1977).
8 Scandinavian Society for Clinical Chemistry and Clinical Physiology: Recommended method for the determination of creatine kinase in blood. Scand. J. clin. Lab. Invest. *36:* 711 (1976).

M. MATHIEU, PhD, Hôpital Debrousse, *F-69322 Lyon Cedex 1* (France)

Enzymes in Health and Disease. Inaug. Scient. Meet. Int. Soc. Clin. Enzymol., London 1977, pp. 103–106 (Karger, Basel 1978)

Computer-Assisted Control of a Clinical Enzymology Laboratory[1]

A. R. HENDERSON

Department of Clinical Biochemistry, University Hospital,
University of Western Ontario, London, Ont.

The operation of the Coulter Status I Laboratory Management System is described as it relates to the processing of Clinical Enzymology data in a 400-bed teaching hospital. The system is based on an extended DEC PDP8e with 24K of core store. It has an IOMEC disc drive (10.24×10^6 bytes on two discs) and magnetic tape unit, 2 ODEC line printers (280 lines per min), 1 label printer, 1 exception printer, 7 cathode ray tube displays and key-boards (CRT) and SMA 6/60 and 12/60 interfaces.

A patient file contains the patient's full name, hospital number (6-digit number), sex, date of birth, bed location and physician's name (there are duplicate checks for name, number and location). This file requires less than 60 keystrokes, and it can be recalled by entering either the name, hospital number or bed location.

Once the patient file has been created, test requests can be made. Tests are entered manually by CRT either individually or by groups (i.e. battery request). Mnemonics of up to four characters are used for CRT entry but the mnemonic lengths have been reduced with staff experience of the system (e.g. lactate dehydrogenase isoenzymes = LDIS→LI). Batteries have been extensively used, i.e. cardiac profile (CAR) = lactate dehydrogenase and creatine kinase with isoenzymes; liver profile (LIV) = transaminases, alkaline phosphatase and γ-glutamyl transpeptidase; transaminases (TR) = alanine and aspartate aminotransferases. On requesting a test on a patient, through the CRT, the sample receives a number (between 0000 and 4095) and all manipulations of the data are subsequently made by means of this number.

[1] Supported by a grant from the Ontario Heart Foundation.

At this stage, the label printer produces small labels for specimen tube labelling. The time of specimen receipt is usually used as the specimen date and time but back or forward timing may be used although this involves more key-strokes. Worksheets for one or more analytes are called when required. In addition to the sample number, they contain the patient's name and hospital number.

Data are produced from LKB 8600 Analyzers with reaction rate calculators or from an ENI Centrifugal Analyzer. Assays are checked by conventional procedures, i.e. recorder tracing, error messages, quality control results, etc., and results are manually entered from the worksheets. Data entry is by a self-prompting CRT response which gives, on entry of the sample number, the patient file to enable complete identification to be made before data entry. The available format for that test is also presented.

These data are then checked for exceeding either the action limit and/or delta checks which is defined as:

$$\frac{\text{Result (previous)} - \text{result (present)}}{\text{Result (previous)}} \cdot 100.$$

The values currently used in the enzyme laboratory are shown in table I. Any value failing either of these checks is printed out on the exception printer and the laboratory supervisor *must* then accept or cancel the data or re-run the analysis. Otherwise, the data will appear on the patient's Cumulative Report as requiring review. Data so caught but accepted are flagged on the report to draw the physician's attention to a likely significant change in the data. Data lying outside the normal range (15 age ranges for both male and female) are likewise flagged.

The action limit check will detect (1) many transcription and entry errors, and (2) moderate or gross pathological change on the first or subsequent specimen analyzed. The limits shown in table I (which can be up to three significant digits, in the test format) were found to be the best compromise between catching too many analyses or allowing too little data to be checked. For any patient up to five analytes can have special action limits applied which overide the test action limit. Examples of their use would be for alkaline phosphatase in cases of biliary cirrhosis or amylase in macroamylaseaemia. Action limits have been variously described in the literature as single parameter checks [3] or alert limits [5] and many computer systems are programmed for this facility [1].

When a second or subsequent analysis is made for the same analyte, then the delta check can be applied. For example, an alkaline phosphatase

Table I. Adult normal ranges (nominal), action limits and delta checks for some serum enzymes

	Normal range U/l[1]		Action limits U/l[1]		Delta check %	
	low	high	low	high	low	high
Alanine transaminase	5	30	5	60	50	50
Aspartate transaminase	10	30	10	60	50	50
Alkaline phosphatase	20	90	20	200	50	50
γ-Glutamyl transpeptidase	0	30	5	100	50	50
Lactate dehydrogenase (LD)	170	320	170	700	50	50
Creatine kinase (CK)	0	135	10	200	50	50
Amylase	70	300	70	450	50	50
CK-1	0	2	0	5	100	100
CK-2	0	4	0	10	50	50
CK-3	24	111	24	300	50	50
LD-1	14	26	10	30	20	20
LD-2	29	39	25	40	20	20
LD-3	20	26	15	30	20	20
LD-4	8	16	5	20	20	20
LD-5	6	16	5	20	20	20

[1] Percentage for LD isoenzymes.

of 110 U/l would fail our current delta check if the previous result had been, say, 70 U/l. This represents a 57% increase and our high delta check (table I) for this enzyme is 50%. However, a *drop* of 40 U/l, from an activity of 110 U/l would pass the low delta check of 50%. Many computer systems use this check [1, 2, 4, 5].

We carry out between 200 and 300 enzyme assays, including isoenzymes (counted as single assays) every weekday. Up to 30% of these assay results are 'caught' by the exception printer as they fail either/both the action limit (80% +) and delta check (less than 20%). For up to 90 results reviewed each day, we find less than 2 to be erroneous, i.e. a total discovered error incidence of about 1%.

While the computer system automatically captures these data following entry, the review process requires the time of supervisory personnel to discover a 1% error incidence! Can a rational approach be made to estimating the limits which maximize the error capture in clinical enzymology? Probability density functions for delta changes have been determined for SMA 6 results at three ranges and two time intervals [4]. It seems likely that a similar approach could be taken for enzyme results both for action limits and delta checks.

References

1 Johnson Associates: Achieving the optimum information system for the laboratory, pp. 77–79 (Johnson Associates, Northbrook 1975).

2 LADENSON, J.H.: Patients as their own controls: use of the computer to identify 'laboratory error'. Clin. Chem. *21:* 1648–1653 (1975).

3 NOSANCHUK, J.S. and GOTTMAN, A.W.: CUMS and delta checks. A systematic approach to quality control. Am. J. clin. Path. *62:* 707–712 (1974).

4 WHEELER, L.A. and SHEINER, L.B.: Delta check tables for technicon SMA 6 continuous-flow analyzer. Clin. Chem. *23:* 216–219 (1977).

5 WHITEHURST, P.; SILVO, T.V. DI, and BOYADJIAN, G.: Evaluation of discrepancies in patients' results – an aspect of computer-assisted quality control. Clin. Chem. *21:* 87–92 (1975).

A.R. HENDERSON, MB, University Hospital, Department of Clinical Biochemistry, *London, Ontario* (Canada)

Enzymes in Health and Disease. Inaug. Scient. Meet. Int. Soc. Clin. Enzymol., London 1977, pp. 107–114 (Karger, Basel 1978)

The Dependence of the Molar Absorptivity of 4-Nitroaniline, 3-Carboxy-4-Nitroaniline and 4-Nitrophenol on the Reaction Conditions

P. LEHMANN, H. LILL, E. SCHAICH and M. GRASSL

Boehringer Mannheim GmbH, Biochemica Werk Tutzing, Forschungszentrum, Tutzing

Introduction

4-Nitroaniline (4-NA), 3-carboxy-4-nitroaniline (3-CA-4-NA) and 4-nitrophenol (4-NP) are frequently used as indicator compounds for photometry in enzymatic analysis, because they allow a direct reaction measurement without deproteinization or any chemical treatment of the cleavage products.

Clinical enzymology has relied upon measurement techniques based upon the absorption coefficient at 405 nm. The most commonly used reference values are listed in table I [1, 2].

Improvements in spectrophotometric instrumentation and standardization techniques, as well as the availability of high-purity synthetic preparations, led to the consideration that the molar absorptivity of these three indicator compounds should be established with greater certainty. Therefore, we have been attempting to evaluate the molar absorption coefficients. Of the many reactions which can be determined, the measurement conditions of three enzyme systems will be discussed (table II) [1, 3].

Methods and Materials

Optical measurements were done on two different Eppendorf Spectrum line photometers (Model 1101 and 1101 M, Eppendorf Gerätebau Netheler & Hinz, Hamburg, Germany), which were equipped with thermostatted cuvette holders. Temperature was determined with Tastotherm P4 thermocouples (Braun Electronic, Frankfurt/Main, Germany), which were calibrated with a reference thermometer (A. H. Thomas, Phila-

Table I

Indicator compound	Linear molar absorption coefficient[1]	References
3-Carboxy-4-nitroaniline (3-CA-4-NA)	9.5	Monotest® γ-GT new (BM)[2]
4-Nitroaniline (4-NA)	9.9	SZASZ *et al.* [1] Monotest® γ-GT and Monotest®-LAP(BM)[2]
	9.62	E. Merck, Darmstadt
	10.60	AB Bofors, s-2160
	10.86	Svendson, Fa. Pentapharm [personal commun., 1944]
4-Nitrophenol (4-NP)	18.5	Test kits acid and alkaline phosphatase (BM)[2]
	18.53	RICHTERICH und GAUTHIER [2]

[1] $\varepsilon_{405\ nm}^{25\ ^\circ C} \times 10^2\ m^2 \times mol^{-1}$

[2] BM = Boehringer Mannheim.

Table II

Enzyme	Substrate	Indicator compound
γ-Glutamyltransferase (γ-GT) EC 2.3.2.2	L-γ-glutamyl-3-carboxy-4-nitroanilide (Glu-3-CA-4-NA)	3-carboxy-4-nitroaniline (3-CA-4-NA)
Leucine arylamidase (LAP) EC 3.4.1	leucine-4-nitroanilide	4-nitroaniline (4-NA)
Alkaline phosphatase (AP) EC 3.1.3.1	4-nitrophenyl phosphate	4-nitrophenol (4-NP)

delphia). The accuracy of the photometers was checked according to the instructions of the manufacturers.

Tables III–V show the final reagent concentrations in the reaction mixture of the three assay systems.

The reagent mixture of each system was prepared in a calibrated flask with freshly distilled deionized water, the pH of the solution was adjusted at 25 °C with a pH-meter calibrated at 25 °C with standardized reference buffers.

Table III. Conditions of the γ-GT assay

Temperature	25 °C
pH (25 °C)	8.25
Tris (hydroxymethyl)-aminomethane	100 mmol × l^{-1}
Glycylglycine	100 mmol × l^{-1}
Glu-3-CA-4-NA	4 mmol × l^{-1}
3-CA-4-NA	6–180 μmol × l^{-1}
Volume fraction of sample	0.09 (1:11)

Principle

$$\text{Glu-3-CA-4-NA} + \text{Gly-gly} \xrightarrow{\gamma\text{-GT}} \text{Glu-gly-gly} + \text{3-CA-4-NA}$$

Table IV. Conditions of the leucine arylamidase assay

Temperature	25 °C
pH (25 °C)	7.50
Tris (hydroxymethyl) aminomethane	50 mmol × l^{-1}
Leucine-4-nitroanilide	4 mmol × l^{-1}
NaCl	250 mmol × l^{-1}
4-Nitroaniline	5–170 μmol × l^{-1}
Volume fraction of sample	0.063 (1:16)

Principle

$$\text{Leu-4-NA} + H_2O \xrightarrow{\text{'LAP'}} \text{Leu} + \text{4-NA}$$

Table V. Conditions of the alkaline phosphatase assay

Temperature	25 °C
pH (25 °C)	9.80
Diethanolamine	1 mol × l^{-1}
$MgCl_2$	0.5 mmol × l^{-1}
4-Nitrophenylphosphate	10 mmol × l^{-1}
4-Nitrophenol	3–100 μmol × l^{-1}
Volume fraction of sample	0.017 (1:60)

Principle

$$\text{4-NPP} + H_2O \rightarrow \text{4-NP} + P_i$$

The exact amount of the indicator compound was accurately diluted at 25 °C to the desired concentrations with doubly distilled water. Before pipetting, the temperature of the reagent solution and of the specimen was adjusted to the calibration temperature of the pipettes. A 2.00-ml portion of the reagent mixture was placed into a glass cuvette (light path 1 cm). When the reagent mixture was equilibrated at 25 °C and a constant absorbance reached, the initial absorbance was read and the absorbance followed for a 5- to 10-min period. Then 200 µl of the specimen was added. The absorbance was read for a 10-min period until a stable end-point was reached. To completely eliminate pipetting errors, the cuvette was weighed before and after each pipetting step.

3-Carboxy-4-nitroaniline was purchased from Ferrak (Berlin, Germany) as free acid and further purified by recrystallization from water/ethanol. 4-Nitroaniline was from Riedel de Haen (Seelze-Hannover, Germany). 4-Nitrophenol was obtained from Fluka (Buchs, Switzerland). L-γ-Glutamyl-3-carboxy-4-nitroanilide, ammonium salt; leucine-4-nitroanilide, crystallized; 4-nitrophenylphosphate, disodium salt and glycylglycine were products of Boehringer Mannheim (Mannheim, Germany). All other chemicals were purchased from E. Merck (Darmstadt, Germany).

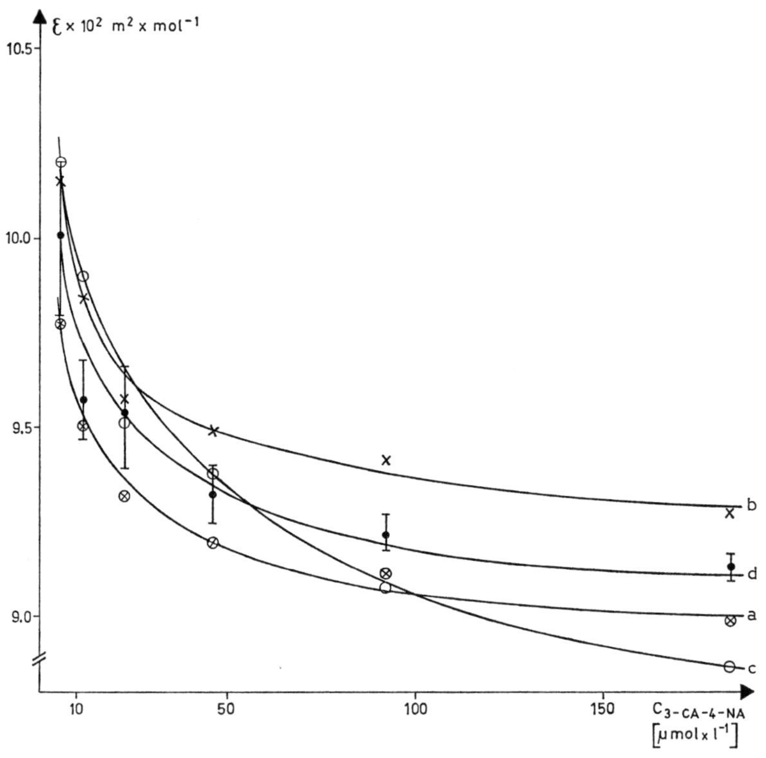

Fig. 1. Effect of 3-CA-4-NA concentration and various components of the assay mixture on the molar absorptivity of 3-CA-4-NA. Sample in: (a) water; (b) Tris buffer, pH = 8.25, 100 mmol × l^{-1} + glycylglycine 100 mmol × l^{-1}; (c) solution b + glu-3-CA-4-NA 4 mmol × l^{-1}; (d) solution c + serum albumin 0.5%. The ranges (curve d) of experimental error are indicated by bars.

Results

Following the procedure outlined above, we obtained at Hg 405 nm and at 25 °C the molar absorptivities summarized in figures 1–3.

3-Carboxy-4-Nitroaniline

As indicated in figure 1, the molar absorptivity is varying dramatically with changes in concentration. Nonlinearity occurs at low indicator compound concentration.

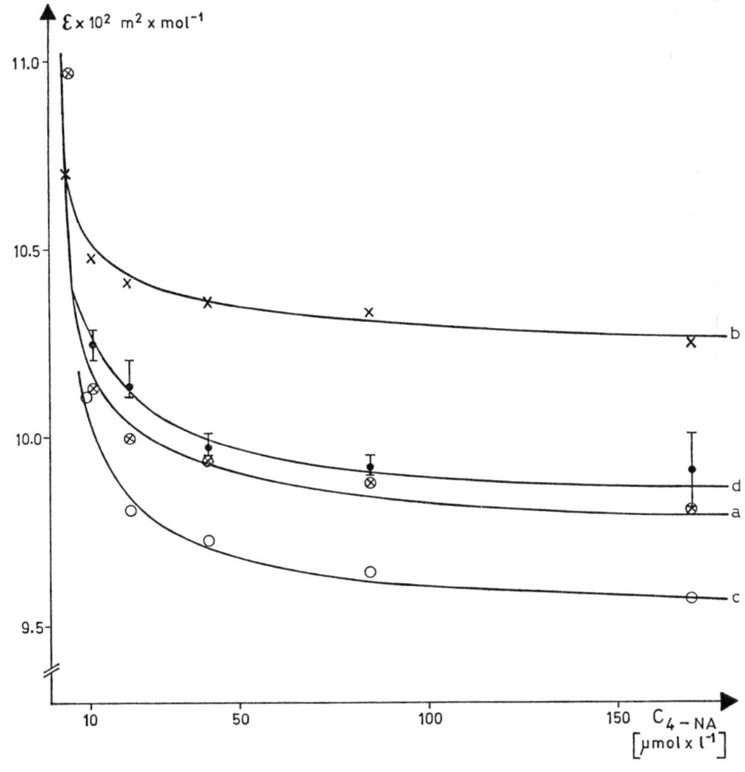

Fig. 2. Effect of 4-nitroaniline concentration and of various components of the assay mixture on the molar absorptivity of 4-nitroaniline. Sample in: (a) water; (b) Tris/HCl, pH=7.5, 50 mmol × l^{-1} + NaCl 250 mmol × l^{-1}; (c) solution b + leucine-4-nitroanilide 4 mmol × l^{-1} + serum albumin 0.5%; (d) solution b + leucine-4-nitroanilide 4 mmol × l^{-1} + serum albumin 1.0%. The ranges of experimental error are indicated by bars.

The experiments further indicate that Tris buffer in the presence of glycylglycine elevates the molar absorptivity (curve b), the presence of the substrate (L-γ-glutamyl-3-carboxy-4-nitroanilide) causes a decrease (curve c), while serum elevates the ε-value (curve d).

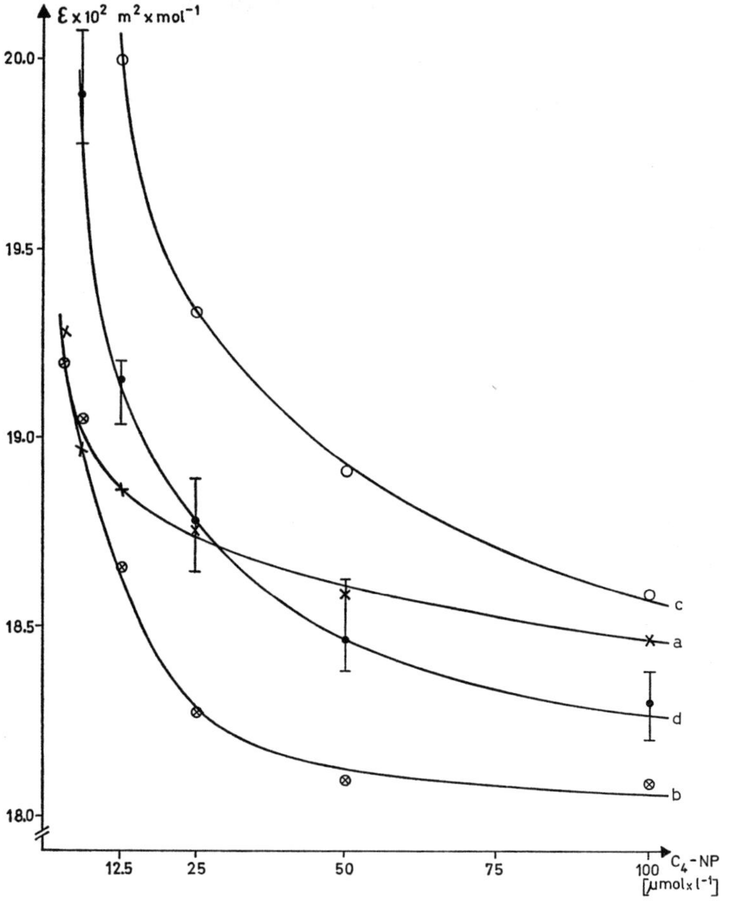

Fig. 3. Effect of 4-nitrophenol concentration and of various components of the assay mixture on the molar absorptivity of 4-nitrophenol. Sample in: (a) diethanolamine buffer, pH=9.8, 1 mol × l^{-1}; (b) solution a+$MgCl_2$ 0.5 mmol × l^{-1}; (c) solution b+*p*-nitrophenylphosphate 10 mmol × l^{-1}; (d) solution c+serum albumin 0.5%. The ranges of experimental error are indicated by bars.

4-Nitroaniline

Nonlinearity of the ε-value occurs at low 4-NA-concentration in a manner similar to that observed for 3-CA-4-NA. Experiments with different additives indicate that in the presence of the univalent electrolyte NaCl, Tris buffer elevates the ε-value (curve b), while the substrate (curve c) and serum albumin (curve d) depress the molar absorptivity (fig. 2).

4-Nitrophenol

The dependence of the ε-value obtained with this compound is comparable with those obtained with 3-CA-4-NA or 4-NA (fig. 3).

Discussion

As presented above, the molar absorptivities of 3-CA-4-NA, 4-nitroaniline and 4-nitrophenol are not constant but depend on the conditions under which they are measured. One might be tempted to conclude that for the molar absorptivity the structure of solvent plays a dominant role.

The interaction of light with material can be related formally to the dipole induced in the material by the oscillating electric field. The absorption of light is proportional to the electric dipole transition moment, which is characteristic for the electronic structure of a molecular group. Therefore, in general, any change will alter the spectrum.

The use of induced dipole interactions to treat optical properties is well established [4–6].

The exact quantitative effect of the dipolar interaction, however, cannot be evaluated with the data available now. A statement would require further experimental investigations and theoretical considerations.

Summary

The ε-values of 3-carboxy-4-nitroaniline, 4-nitroaniline and 4-nitrophenol at Hg 405 nm and 25 °C were redetermined. In the case of 3-carboxy-4-nitroaniline the kinetic test of γ-glutamyltransferase based on the principles outlined by Szasz et al. was used.

The ε-values of 4-nitroaniline and 4-nitrophenol were determined under the conditions of the leucine arylamidase assay and the alkaline phosphatase assay according to the recommendations of the German Society for Clinical Chemistry.

We found that the molar absorptivities of these three indicator compounds are not constant but depend on the conditions under which they are measured. They are strongly influenced by their concentration and by the ionic strength of the solution.

References

1 SZASZ, G.: Selected Method of Clinical Chemistry: Reaction-rate method for γ-glutamyltransferase activity in serum. Clin. Chem. *22:* 2051 (1976).
2 RICHTERICH, R. und GAUTHIER, E.: Ultramikromethoden im klinischen Laboratorium. Schweiz. med. Wschr. *92:* 1496 (1962).
3 Empfehlungen der Deutschen Gesellschaft für Klinische Chemie. Z. klin. Chem. klin. Biochem. *8:* 658–660 (1970).
4 LIPTAY, W.: Elektrochemie-Solvatochromie. Angew. Chem. *81:* 195 (1969).
5 MURRELL, J.N.: Elektronenspektren organischer Moleküle, pp. 257–268 (Bibliographisches Institut AG, Mannheim 1967).
6 KORTÜM, G.: Lehrbuch der Elektrochemie, pp. 148–152 (Verlag Chemie, Weinheim 1966).

P. LEHMANN, PhD, Boehringer Mannheim GmbH, Biochemica Werk Tutzing, Bahnhofstrasse 9–15, *D-8132 Tutzing/Obb. (FRG)*

Enzyme Methods

Enzymes in Health and Disease. Inaug. Scient. Meet. Int. Soc. Clin. Enzymol., London 1977, pp. 115–117 (Karger, Basel 1978)

Enzyme Multiplied Immunoassay of Serum Digoxin Using the LKB 2074 Calculating Absorptiometer

ANNE M. DUNCAN and ALAN BELFIELD[1]

Department of Biochemistry, Royal Infirmary, Glasgow

Introduction

The measurement of serum digoxin concentration is of considerable help to the clinician during digitalization of patients. To minimize toxicity, a need exists for a simple digoxin assay suitable both for routine and emergency use. For this reason, we investigated the enzyme multiplied immunoassay technique (EMIT) for the analysis of serum digoxin using the LKB 2074 Calculating Absorptiometer.

Materials and Methods

Enzyme multiplied immunoassay of digoxin. Reagent kits and calibration standards, manufactured by Syva Corporation, Palo Alto, Calif., were obtained from Gilford Instruments Ltd., Teddington, Middlesex. The kit contents were reconstituted according to the manufacturer's instructions and working reagents prepared as directed. Each calibrator and test serum was analyzed in duplicate and absorbance measured against water as a reference. Two LKB racks were employed, the first with cuvettes containing the calibrators, ranging in value from 0 to 8.96 nmol digoxin per litre, followed by the four test sera; in the second rack, the order of the cuvettes was reversed. Calibrators and sera (100 µl) were pipetted into the appropriate cuvettes, pretreatment reagent (25 µl) added and the racks vortex mixed for 10 sec. After 5 min, 0.625 ml of the combined antidigoxin and substrate reagent was dispensed into each cuvette and the racks again vortexed before being transferred to a heating block at 30 °C. After 15 min, 250 µl of starting reagent (digoxin-enzyme conjugate) at 30 °C was added and the racks mixed. Immediately after

[1] We thank Mrs. MORAG ALLEN for her help with the radioimmunoassay.

this addition, the absorbance of each cuvette was measured at 340 nm in the LKB 2074 Calculating Absorptiometer (LKB Produkter-AB BROMMA, Sweden). After exactly 30 min incubation at 30 °C the absorbance was again measured. Mean differences between the initial and final readings were calculated, and the values for the test sera obtained from a calibration line drawn as specified on the graph paper provided.

Radioimmunoassay of digoxin. This was performed using 'Immophase' kits (Corning Medical Diagnostics, Medfield, Mass.). Quality control sera, used in both the radioimmunoassay and EMIT, were also obtained from this manufacturer.

Investigational procedure. 35 sera from patients receiving digoxin were assayed by both methods. To test for non-specific reactions, ten samples from subjects not receiving digoxin medication were analyzed by EMIT. The recovery of the method was investigated by analysis of pooled serum to which known amounts of digoxin had been added. To assess precision and accuracy, ten replicate measurements of digoxin concentration were made on each of two quality control sera A and B.

Results

The digoxin concentrations of the 35 sera ranged from 0.9 to 6.1 nmol/l when analyzed by radioimmunoassay. The slope of the line of regression of the EMIT upon radioimmunoassay results was 0.93, standard error of the mean (SEM) 0.09 and intercept 0.18, SEM 0.27. A correlation coefficient of 0.87 ($p<0.01$) was obtained. The paired difference *t*-test showed no difference between results using the two methods of analysis. None of the 10 samples from patients not receiving digoxin gave a result significantly different from zero. Recoveries of digoxin added to serum were carried out at three concentrations: 1.28, 2.56 and 5.12 nmol/l. These three sera were assayed in duplicate on each of four days and recoveries of 1.45, SEM 0.06, 2.79, SEM 0.07, and 5.67, SEM 0.09 nmol/l obtained. Quality control serum A was found to have a mean concentration of 1.75 nmol/l, compared with a manufacturer's stated value of 1.92 nmol/l, standard deviation 0.18 nmol/l; for serum B, the mean was 4.10 nmol/l, standard deviation 0.52 nmol/l, stated value 3.84 nmol/l.

Discussion

Our preliminary work was done using a Gilford 300N spectrophotometer. Like ROSENTHAL *et al.* [1], we experienced difficulty in attaining satisfactory precision due in part to the susceptibility of the flow-cell to small air bubbles. The EMIT results of the present study were obtained using an LKB 2074 Calculating Absorptiometer which possesses several advantages over instruments dependent on a flow-through cell: the assay is carried out in the same disposable cuvette through which initial and final absorbances are measured, thus eliminating problems associated with flow-cells and requiring half of the volume of reagents recommended by the manufacturer; the cuvettes are read automatically at short, precise time intervals in metal racks which can be held in the heating block during incubation.

The precision of the assay depends on the accurate measurement of very small differences between two relatively large absorbances. To eliminate any slight bias in timing, the order of the calibrators and test samples was reversed in the duplicate rack. The overall precision of the EMIT, coefficient of variation 10–15%, is similar to that quoted by SUN and SPIEHLER [2] (7.8–13.9%), but worse than that of ROSENTHAL *et al.*, who made considerable effort to remove all particulate matter from their reagents and obtained values ranging from 5.5 to 8.4%. We found recoveries of digoxin by EMIT to range from 109 to 113%. Both SUN and SPIEHLER and ROSENTHAL *et al.* obtained recoveries greater than 100%.

In summary, we found the LKB 2074 Calculating Absorptiometer to be a suitable instrument for the analysis of digoxin by EMIT.

References

1 ROSENTHAL, A.F.; VARGAS, M.G., and KLASS, C.S.: Evaluation of enzyme-multiplied immunoassay technique (EMIT) for determination of serum digoxin. Clin. Chem. *22:* 1899–1902 (1976).

2 SUN, L. and SPIEHLER, V.: Radioimmunoassay and enzyme immunoassay compared for determination of digoxin. Clin. Chem. *22:* 2029–2031 (1976).

A.M. DUNCAN, BSc, Department of Biochemistry, Royal Infirmary, *Glasgow G4 0SF* (Scotland)

Enzymes in Health and Disease. Inaug. Scient. Meet. Int. Soc. Clin. Enzymol., London 1977, pp. 118–122 (Karger, Basel 1978)

A Sensitive and Simple Method for the Determination of Aspartate Aminotransferase Isoenzymes

KAARINA OJALA and AARNE KONTTINEN

Kivelä Municipal Hospital, Helsinki

Introduction

Aspartate aminotransferase (AST, EC 2.6.1.1) is known to consist of two isoenzymes. These are found in various human tissues without any proven organ-specific distribution. Their localization is characterized by their distribution within the cell. The anionic isoenzyme is localized in the cytoplasm (c-AST, AST I) and the cationic one in the mitochondria (m-AST, AST II). Thus, the significance of m-AST isoenzyme in serum is as an indicator of damage to the double membrane of the mitochondria.

So far, the clinical use of the AST isoenzyme determination has been limited by insensitivity or complexity of the methods. The most sensitive electrophoretic method, presented by BOYDE [1], is semiquantitative, based on the visual inspection of vanishing fluorescence. The first electrophoretic method allowing the quantitation of AST isoenzymes in serum was published by MURROS *et al.* [3] 1972. In this method, the electrophoretic run is carried out on an agarose plate. Its sensitivity is about 6 U/l (25 °C), which is not enough for clinical use.

The purpose of the present work was to develop a sensitive and simple method suitable for routine use. The method is based on the electrophoretic separation of the isoenzymes on cellulose acetate and visualization of the fractions by colour reaction with Fast Violet B.

Materials and Methods

For the assay of total AST activity (kinetic UV-test, 25°C), the commercial test kit of KABI was used, the normal upper limit being 20 U/l.

Beckman Microzone system with Beckman cellulose acetate membranes and barbital buffer of 0.075 mol/l, pH 8.6 was used for electrophoresis.

The agar used was Difco Agar Noble. Pyridoxal phosphate (PP) was purchased from KABI in tablet form (0.83 mg/tablet). Fast Violet B salt was purchased from Sigma and L-aspartic acid and 2-oxoglutaric acid from Merck.

The stock agar solution (4%) is prepared by melting down 2 g agar in 50 ml water in a boiling water bath. The solution is divided into 4-ml fractions in tubes and can be kept for about 2 weeks at 4 °C.

Substrate buffer is prepared as follows: Tris (hydroxymethyl) aminomethane 1.2 g, L-aspartic acid 2.66 g, 2-oxoglutaric acid 350 mg, Na-EDTA 460 mg and polyvinyl-pyrrolidone 3 g are dissolved in redistilled water, adjusting pH to 7.4 and volume up to 100 ml. Store at 4 °C, which will provide stability for at least 1 month.

Substrate gel. One tube of the stock agar solution is melted down in a boiling water bath and then mixed with 4 ml substrate buffer and 0.25 ml pyridoxalphosphate solution, prepared by dissolving one tablet of PP in 1 ml water. After mixing, the solution is poured immediately into a Petri dish (diameter about 10 cm).

In final form, the reagents are as follows: Tris 50 mmol/l, L-aspartate 100 mmol/l, 2-oxoglutarate 12 mmol/l, Na-EDTA 5 mmol/l and PP 0.12 mmol/l.

PP solution is prepared just before use; the substrate gel can be stored for some days at 4 °C.

Electrophoresis. Sample volumes of 0.25–1.00 µl, depending on AST activity, are applied on cellulose acetate strips and electrophoresed in barbital buffer for 20 min at 250 V.

After electrophoresis, the membrane is layered onto the substrate gel and incubated at 37 °C for 30 min. The strip is not removed from the gel after this time, but layered with another strip, which has been soaked in the colour solution (Fast Violet B 40 mg/ml H_2O, prepared just before use) and blotted slightly. The incubation is continued for a further 10 min at 37 °C.

The coloured electrophoresis membrane is then fixed in 5% acetic acid for about 2 min, rinsed in water and air-dried. The dry strip is scanned at 540 nm with Helena Auto Scanner.

Results

Electrophoretic separation of AST isoenzymes is illustrated in fig. 1.

For testing the linearity of the reaction, a serum sample with AST activity of 60 U/l was analyzed by using sample volumes of 0.25–1.00 µl and incubation times of 10, 20 and 30 min. The results are shown in figure 2.

The reaction seems to proceed fairly linearly up to a total activity of about 50 U/l when a sample volume of 1 µl and incubation time of 30 min are used. With higher activities, up to 200 U/l, sample volumes of less than 1 µl have to be used. If the activity exceeds 200 U/l, it is necessary to make a dilution. The sensitivity of the method is about 1 U/l.

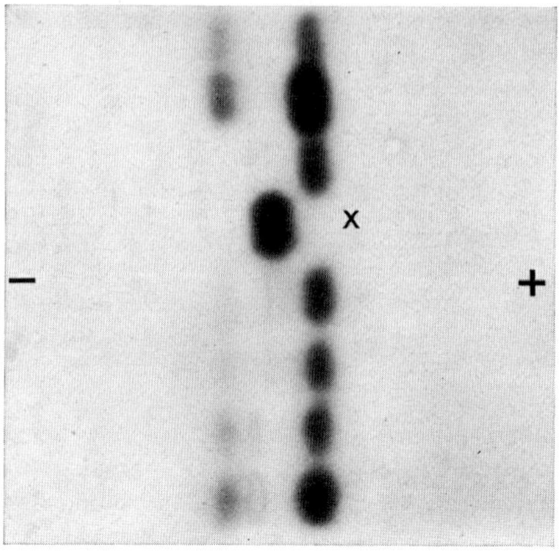

Fig. 1. Electrophoretic separation of AST isoenzymes. The sample marked with X is a complex of IgG and AST (see text).

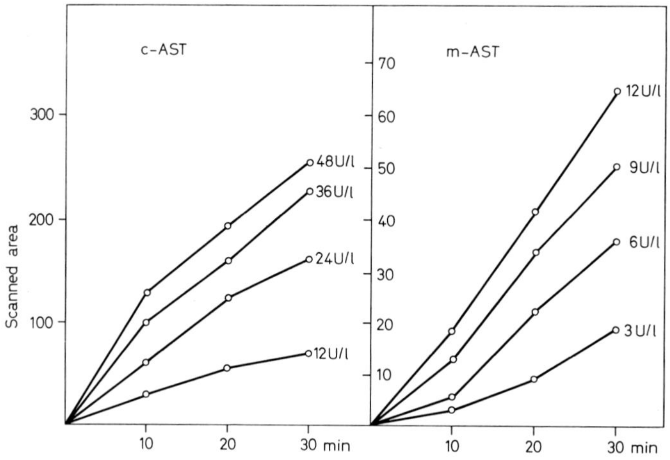

Fig. 2. The scanned areas of the isoenzymes c-AST and m-AST plotted against the incubation time.

Conjugated bilirubin interferes with the test by reacting directly with the diazonium salt; in highly icteric serum, an artefact fraction close to the application point is formed.

Comments

The proposed method is easily adapted for routine use. The procedure is rapid, requiring not more than 1 h to perform. Cellulose acetate membranes and barbital buffer are used for protein electrophoresis in most clinical laboratories. Almost all reagents are fairly stable and can be prepared beforehand. The sensitivity of the isoenzyme detection is increased by using optimal reagent concentrations.

The two-phase incubation and the fixing procedure described result in discolouration of the background which thereby clearly improves the accuracy and sensitivity of the method.

Conjugated bilirubin causes a stained fraction in electrophoresis, but its location does not disturb the calculation of the value of the isoenzyme

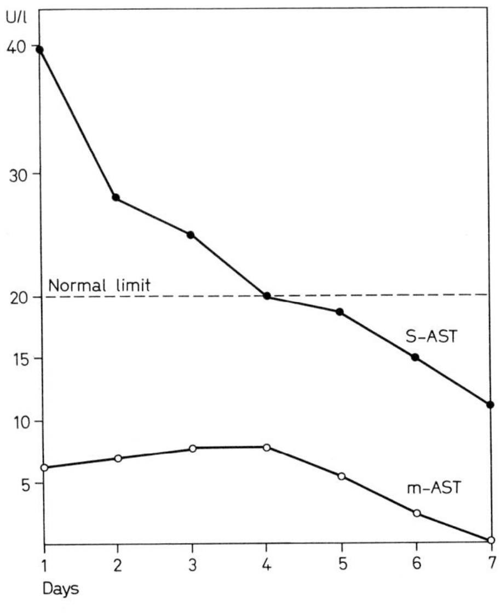

Fig. 3. Total S-AST and m-AST performed on a patient with myocardial infarction during 7 consecutive days.

activity. Indeed, at the same position we found serum AST activity bound abnormally with IgG (fig. 1, marked with X). This rare observation was confirmed by immunological precipitations and is similar to that seen in macroamylasaemia.

The mitochondrial AST isoenzyme was not found in the serum of 25 healthy people studied. This is in conformity with other observations as reviewed by WILKINSON [5], but at variance with the findings of BOYDE [1], who, from results generated by his semiquantitative method, suggested that there might be a mitochondrial AST in normal serum.

Mitochondrial AST isoenzyme is augmented in acute myocardial infarction. Furthermore, the pathological activity appears to last several days longer than the elevation of total AST activity (fig. 3) in conformity with earlier observations using less sensitive methods [2, 4].

Summary

A sensitive and very simple method for the determination of aspartate aminotransferase isoenzymes, based on cellulose acetate electrophoresis, is described. Mitochondrial isoenzyme activities down to about 1 U/l (25 °C) can be detected. Only cytoplasmic AST was found in 25 normal sera analyzed. After the onset of acute myocardial infarction, mitochondrial AST activity remains pathological in the serum several days longer than total AST.

References

1 BOYDE, T.R.C.: Detection and assay of mitochondrial aspartate aminotransferase in serum. Z. klin. Chem. klin. Biochem. *6:* 431–435 (1968).

2 BOYDE, T.R.C.: Serum levels of the mitochondrial isoenzyme of aspartate aminotransferase in myocardial infarction and muscular dystrophy. Enzym. biol. clin. *9:* 385–389 (1968).

3 MURROS, J.; KONTTINEN, A., and SOMER, H.: An electrophoretic method for the quantitation of aspartate aminotransferase isoenzymes. Clin. chim. Acta *41:* 263–268 (1972).

4 MURROS, J.; KONTTINEN, A., and SOMER, H.: Mitochondrial aspartate aminotransferase in myocardial infarction. Clin. chim. Acta *48:* 241–242 (1973).

5 WILKINSON, J.H.: Diagnostic enzymology, p. 92 (Arnold, London 1976).

K. OJALA, MSc, Kivelä-Hesperia Hospitals, Central Laboratory, Sibeliuksenkatu 12–14, *SF-00260 Helsinki 26* (Finland)

Enzymes in Health and Disease. Inaug. Scient. Meet. Int. Soc. Clin. Enzymol., London 1977, pp. 123–126 (Karger, Basel 1978)

Pancreatic Lipase and Its Inhibitors

J. Prokopowicz and M. Rublewska
Department of Clinical Diagnostics, Medical School, Białystok

Introduction

The determination of pancreatic lipase activity is still important in diagnosis of pancreatic diseases [3]. Increased activity of this enzyme in serum is more prolonged than amylase [5]. No data are available concerning activators and inhibitors of pancreatic lipase. Considerable interest has been focused on the physico-chemical properties of human pancreatic lipase (triacylglycerol lipase, EC 3.1.1.3) from pancreatic juice [1, 4, 6]. The purpose of the current studies was to clarify some physico-chemical properties of human serum pancreatic lipase from controls and patients with acute pancreatitis and to investigate possible inhibitors of this enzyme in serum.

Material and Methods

Sera. Blood was collected from 35 patients with acute pancreatitis and from 20 controls.

Lipase assay. Lipase activity was measured according to Cherry and Crandall [2].

Protein determination. Protein concentration was determined spectrophotometrically at 260 and 280 nm, according to formula protein concentration mg/ml = 1.45 A_{280} nm −0.74 A_{260} nm.

Chromatographic studies. Sephadex G-200 columns (40 × 2.5 cm), were equilibrated with 0.5 M NaCl and eluted with the same solution. DEAE Sephadex A-50 columns (15 × 2.5 cm) were eluted with a linear concentration gradient of NaCl (0–1.0 M).

Activity determination of pancreatic lipase inhibitors. Standard solution of hog pancreatic lipase (Sigma) 180 mIU/ml was incubated for 30 min at 37 °C in water bath with eluates. The difference between initial and final activity, expressed in mIU/ml, was arbitrally accepted as inhibitory activity.

1

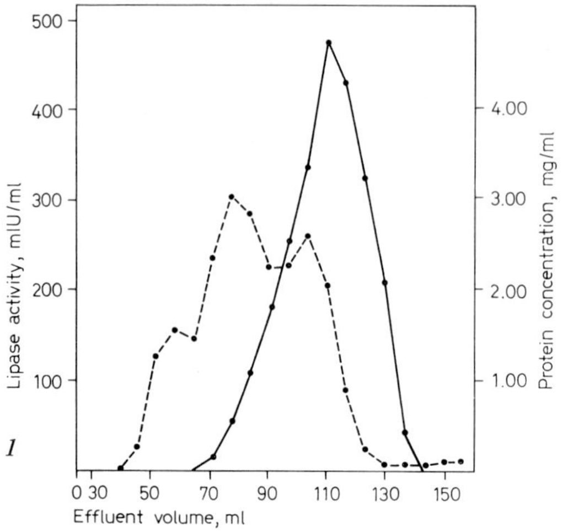

2

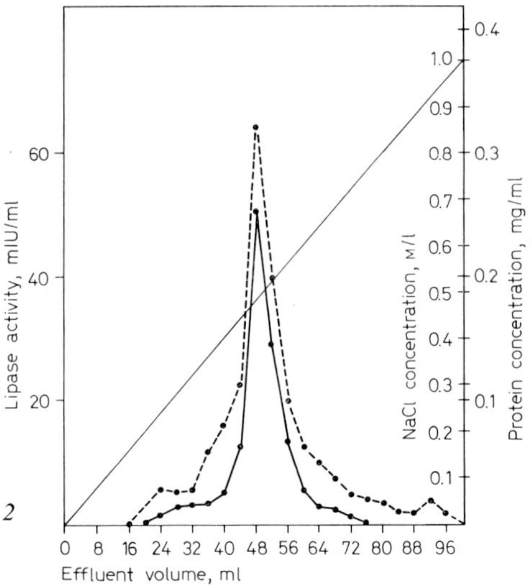

Fig. 1. Gel filtration on Sephadex G-200 of serum with high activity pancreatic lipase. Serum from patient with acute pancreatitis, enzyme activity 2,500 mIU/ml. —— = Activity of pancreatic lipase; – – – = protein concentration.

Fig. 2. Rechromatography on DEAE Sephadex A-50 of serum pancreatic lipase partially purified on Sephadex G-200. The same sample of serum as used in figure 1. —— = Activity of pancreatic lipase; – – – = protein concentration.

Results

As shown in figure 1, pancreatic lipase is localized in one peak. The same distribution of this enzyme was obtained when normal serum was tested.

Figure 2 illustrates that after this procedure pancreatic lipase is still localized in one peak. The highest activity is eluted at 0.5 M NaCl. After rechromatography, the pancreatic lipase is purified about 150-fold in comparison to serum. Using sieve chromatography (Sephadex G-200) the molecular weight of pancreatic lipase was estimated to be about 45,000. Maximum activity of pancreatic lipase was found at pH 8–9. This enzyme is thermolabile. It has been found that incubating pancreatic lipase after chromatography on Sephadex G-200 with trypsin (final concentration in tested system 1 mg/ml) did not significantly change the activity of this enzyme. We have found the highest activity is obtainable in the presence of sodium ions at a concentration about 196 mM.

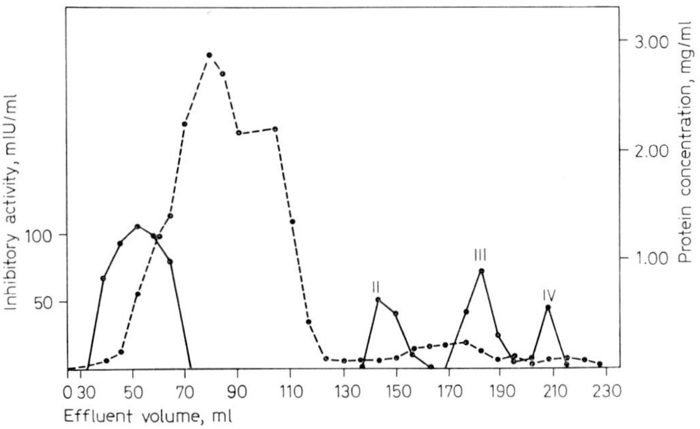

Fig. 3. Gel filtration on Sephadex G-200 of serum with high activity of pancreatic lipase. Serum from patient with acute pancreatitis, enzyme activity 1,600 mIU/ml. —— = Inhibitory activity; - - - = protein concentration.

Figure 3 demonstrates the distribution of inhibitors against pancreatic lipase. In this case, we have observed four peaks. But in some samples of serum we found only two peaks. The molecular weight of these proteins ranged from 13,000 to 400,000. According to our experiments they are progressive inhibitors.

Discussion

Both in controls and in patients with acute pancreatitis, we have found only one serum lipase. Our data are in line with Kimura *et al.* [47], who found two lipases in pancreatic juice but identical immunologically. The difference between both lipases depends on the presence of a phospholipid compound. Recently, De Caro *et al.* [1] found two lipases in human pancreatic juice with small differences between them only in isoelectric points. The molecular weight determined by us is in agreement with other authors [6]. It is interesting to note that serum pancreatic lipase from patients with acute pancreatitis is more sensitive to temperature inactivation than that of the controls. The number of the pancreatic lipase inhibitors is not related to the activity of this enzyme and may be an individual attribute of the tested person. In controls and in patients with moderately increased pancreatic lipase activity, total inhibitory activity is higher than total pancreatic lipase activity.

Summary

Human pancreatic lipase and its inhibitors were separated from sera obtained from controls and patients with acute pancreatitis. Lipase from both types of sera was localized in one peak and its molecular weight is about 45,000. Maximum activity was found at pH 8–9. Sodium ions enhance activity of lipase. The purified pancreatic lipase is only slightly digested by trypsin. We have found two or even four progressive inhibitors of pancreatic lipase with molecular weights ranging from 13,000 to 400,000.

References

1 De Caro, A.; Figarella, C.; Amic, J.; Michel, R., and Guy, O.: Human pancreatic lipase: a glycoprotein. Biochim. biophys. Acta *490:* 411–419 (1977).
2 Cherry, I.S. and Crandall, L.A., jr.: The specificity of pancreatic lipase: its appearance in the blood after pancreatic injury. Am. J. Physiol. *100:* 266–273 (1932).
3 Katz, S.: Characteristics of fat-splitting enzymes in health and disease. Am. J. Gastroent., N.Y. *27:* 479–487 (1957).
4 Kimura, H.; Kitamura, T., and Tsuji, M.: Studies on human pancreatic lipase. I. Interconversion between low and high molecular forms of human pancreatic lipase. Biochim. biophys. Acta *270:* 307–316 (1972).
5 Tietz, N.W.; Borden, T., and Stepleton, J.D.: An improved method for the determination of lipase in serum. Am. J. clin. Path. *31:* 148–154 (1959).
6 Vandermeers, A.; Vandermeers-Piret, M.C.; Rathè, J., and Christophe, J.: On human pancreatic triacylglycerol lipase: isolation and some properties. Biochim. biophys. Acta *370:* 257–268 (1974).

J. Prokopowicz, MD, Department of Clinical Diagnostics, Medical School, *PL-15-276 Bialystok* (Poland)

Enzymes in Health and Disease. Inaug. Scient. Meet. Int. Soc. Clin. Enzymol., London 1977, pp. 127–130 (Karger, Basel 1978)

Quantitative Determination of Activities of Alkaline Phosphatase Isoenzymes in Serum

KATRINE B. WHITAKER, L. G. WHITBY and D. W. MOSS

Royal Postgraduate Medical School, London

Introduction

The value of characterization of alkaline phosphatase isoenzymes is well established [1], mainly as a result of qualitative assessment of electrophoretic separation of the isoenzymes in serum. This method is adequate for many diagnostic problems. However, in order to realise the full value of isoenzyme studies in diagnosis, quantitative methods are desirable. The majority of clinical problems require the separate measurement of the bone and liver isoenzymes of alkaline phosphatase. Densitometry of electrophoretically separated alkaline phosphatase zones does not fully meet this need, because of the incomplete separation of the bone and liver components. However, the differential inactivation of bone and liver alkaline phosphatase by heat [2] can be adapted to quantitation of the two isoenzymes, and we now describe the routine diagnostic use of such a method.

Methods

The procedure of MOSS and WHITBY [3] was applied to serum specimens submitted for diagnostic investigation and from normal subjects. Total alkaline phosphatase activity and residual activities after incubation for 15 and 25 min at 56 ± 0.05 °C were determined at 37 °C [4].

Results and Discussion

Reference ranges established for apparently healthy people show that adults have, on average, approximately equal activities of liver and bone

Table I. Primary reasons for requesting characterization of serum alkaline phosphatase isoenzymes in 200 consecutive specimens

1. Identification of tissue of origin of an unexplained rise in total serum alkaline phosphatase activity	104
2. Confirmation of the identity of a strongly suspected tissue of origin of an abnormal alkaline phosphatase activity	22
3. Provisional diagnosis which involves the possibility of more than one tissue source of abnormal alkaline phosphatase activity	32
4. To follow changes in a known abnormality of alkaline phosphatase isoenzyme composition	34
5. Determination of isoenzyme composition in presence of a physiologically raised isoenzyme (e.g. from placenta or normally growing bone)	8
Total	200

phosphatases. Liver phosphatase activity rises somewhat with age and accounts for the slight increase in total activity in older age groups, whereas greater bone phosphatase activity explains the higher total activity observed in younger people [4].

The need for quantitative rather than qualitative methods of isoenzyme characterization in diagnosis can be seen by considering the primary reasons for requesting studies of serum alkaline phosphatase isoenzymes (table I). Qualitative methods are probably adequate for most of the cases in category 1, because usually only one isoenzyme is affected. However, even within this group, quantitative methods are desirable where the total alkaline phosphatase activity is close to the upper limit of normal, when qualitative methods are insufficiently sensitive. In a series of 550 requests, 33 total alkaline phosphatase activities were near the upper border of the normal range. Of these, 26 sera were found to have an elevated activity of one of the isoenzymes. Furthermore, 18 sera with total activities well within normal limits had an elevation of liver or bone phosphatase (fig. 1). Where confirmation of the strongly suspected presence of a particular isoenzyme is required (category 2), a qualitative method is also normally adequate. Nevertheless, the quantitative method may reveal an unsuspected abnormal activity of a second isoenzyme. This was true for 5 out of 22 cases in this category.

Quantitative methods of isoenzyme analysis are even more useful in the investigation of cases falling into category 3. These include metastatic malignant disease, where the detection of small increases in the activity of a

1

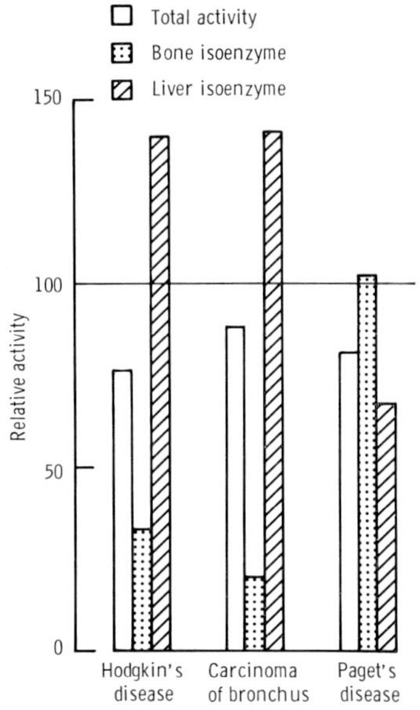

2

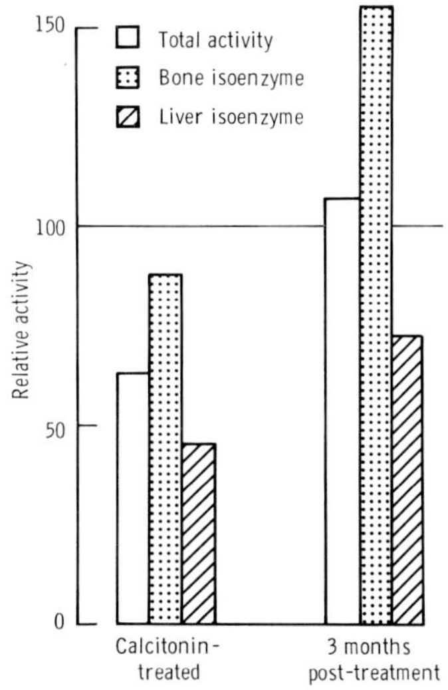

Fig. 1. Total activity of alkaline phosphatase and activities of the bone and liver isoenzymes in serum (expressed as percentages of the corresponding upper limits of normal adjusted for age) in 3 patients in whom total activities were normal but isoenzyme distributions were abnormal.

Fig. 2. Total activity of alkaline phosphatase and activities of the bone and liver isoenzymes (expressed as percentages of the corresponding upper limits of normal adjusted for age) in serum from a 64-year-old woman with Paget's disease, during treatment with calcitonin and 3 months after treatment ceased.

specific isoenzyme can lead to the early discovery of new secondary deposits. Where there are pre-existing metastases already contributing markedly to the total alkaline phosphatase, small changes in a second isoenzyme activity can only be detected by quantitative methods.

Quantitative analysis is virtually essential for requests comprised by categories 4 and 5. Small changes in isoenzyme distribution during the progress or treatment of disease developing over a considerable period are difficult to estimate by electrophoresis because of between-run variability in separation and staining of isoenzyme zones. For example, relatively small changes in the activity of bone phosphatase in serum are useful in monitoring treatment of Paget's disease with calcitonin (fig. 2).

Since visual assessment of changes in a minor isoenzymic component on electrophoretic patterns is particularly difficult, quantitative methods are required to estimate the liver isoenzyme in children, in whom bone phosphatase predominates. Assessment of the contribution of placental phosphatase by electrophoresis is not reliable because its mobility is variable and usually similar to that of the liver and bone isoenzymes. However, the presence of placental-type isoenzymes can be detected in the heat-inactivation method at levels as low as 2% of total activity, because of their effect on the apparent half-life of the liver isoenzyme. Intestinal alkaline phosphatase is similar to the liver isoenzyme in stability, so that its contribution to total activity cannot be determined by the heat-inactivation procedure alone and additional criteria, such as zone electrophoresis, are needed.

References

1 MOSS, D.W.: Alkaline phosphatase isoenzymes. Technical and clinical aspects. Enzyme *20:* 20–34 (1975).
2 MOSS, D.W. and KING, E.J.: Properties of alkaline phosphatase fractions separated by starch-gel electrophoresis. Biochem. J. *84:* 192–195 (1962).
3 MOSS, D.W. and WHITBY, L.G.: A simplified heat-inactivation method for investigating alkaline phosphatase isoenzymes in serum. Clin. chim. Acta *61:* 63–71 (1975).
4 WHITAKER, K.B.; WHITBY, L.G., and MOSS, D.W.: Activities of bone and liver alkaline phosphatases in serum in health and disease. Clin. chim. Acta *80:* 209–220 (1977).

Prof. D.W. MOSS, Department of Chemical Pathology, Royal Postgraduate Medical School, Ducane Road, *London W12 0HS* (England)

Enzymes in Health and Disease. Inaug. Scient. Meet. Int. Soc. Clin. Enzymol., London 1977, pp. 131–136 (Karger, Basel 1978)

Kinetic Enzymatic Method for Automated Determination of Uric Acid in Serum

JOACHIM ZIEGENHORN, MAX BRANDHUBER and KNUT BARTL

Boehringer Mannheim GmbH, Biochemica Werk Tutzing, Forschungszentrum, Tutzing

Introduction

For the enzymatic determination of uric acid in serum, the UV method according to PRAETORIUS *et al.* [2] or the colorimetric method according to KAGEYAMA [2], are commonly used. Both procedures are unsuited for many of the commercially available automated instruments because they require special photometric equipment or a long reaction time. This situation prompted us to develop a method for fast routine analysis of uric acid in serum with use of automated systems.

The new method was based upon the specific degradation of uric acid by uricase (EC 1.7.3.3) and quantification of the resulting H_2O_2 with catalase (EC 1.11.1.6) and aldehyde dehydrogenase (EC 1.2.1.5) according to HAECKEL [1]. Through addition of the competitive inhibitor xanthine the Michaelis constant of uricase was apparently increased to such an extent that the uricase reaction was pseudo-first order in uric acid concentration over a wide range of concentration. Moreover, this reaction became the rate-limiting step of the whole reaction sequence. These facts allowed the specific measurement of uric acid concentrations up to 1.3 mmol/l of serum by means of kinetic fixed-time procedures [3]. The method has been adapted to the Abbott ABA-100 analyzer, the ENI GEMSAEC centrifugal fast analyzer, and the LKB reaction rate analyzer.

Materials and Methods

Apparatus

The automated instruments used were an ENI GEMSAEC centrifugal fast analyzer (Electro-Nucleonics, 368 Passaic Ave., Fairfield, N.J.) including DEC-PDP 8 computer and DECtape (Digital Equipment, Maynard, Mass.), an Abbott ABA-100 analyzer (Abbott Scientific Products Division, South Pasadena, Calif.), and a LKB 8600 reaction rate analyzer (LKB-Produkter, 16125 Bromma 1, Sweden). The manual assays were carried out on Model 1101 M spectrum line photometers (Eppendorf Gerätebau Netheler & Hinz, Hamburg, Germany).

Reagents

Test kit 'Automatenpackung Harnsäure', cat. No. 242616, and standard 'Precimat Harnsäure' (357 μmol/l), cat. No. 164933, were from Boehringer Mannheim, Mannheim, Germany. From the constituents of the test kit a single reagent was prepared which contained pyrophosphate buffer (50 mmol/l, pH 8.5), NADP (1.1 mmol/l), xanthine (210 μmol/l), uricase from hog liver (0.6 kU/l), catalase from beef liver ($\geqq$1 MU/l), and aldehyde dehydrogenase from yeast ($\geqq$1 kU/l). With the LKB 8600, the uricase component of that reagent was added separately as starting agent. The reagents were stable for one day at 25 °C.

Procedures

GEMSAEC: 50 μl of calibration standard or serum sample, dispensed with 200 μl of physiological saline, was pipetted into cavity B of the transfer disc. 500 μl of reagent was pipetted into cavity C of the transfer disc. The disc was placed into the GEMSAEC analyzer which was operated at a wavelength of 340 nm and at 25 °C. The first absorbance reading was taken 35 sec after the start of the reaction (IR[1] = 10), the second reading 150 sec later. The sample ring was loaded in the following way: position 1, distilled water; position 2, standard; and positions 3–16, serum samples. For the calculation of the results, the computer program TC = 5 was used.

ABA-100: 5 μl of calibration standard or serum sample and 250 μl of reagent were pipetted into the multicuvette of the analyzer. By use of the light filter 340 nm/380 nm, the first absorbance reading was taken at 20 sec after the initiation of the run. The second reading was taken 280 sec later. The incubation temperature was 25 °C. The sample ring was loaded in the following sequence: positions 0–2 remained empty (FRR[1]); position 3–4, distilled water; position 5, standard; and positions 6–31 serum samples.

LKB 8600: 50 μl of calibration standard or serum sample and 1 ml of reagent were pipetted with the LKB diluter into the cuvettes of a rack. The rack was placed into the analyzer operating at a wavelength of 340 nm and at 25 °C. The first absorbance reading was taken 10 sec after the addition of 50 μl of uricase solution, the second reading 170 sec later. The first rack was loaded in the following sequence: position 1, distilled water; position 2, standard; and positions 3–10, serum samples.

With all three versions of the kinetic method, there was no need for reagent or sample blanks.

[1] IR and FRR represent 'initial reading' and 'first revolution reading', respectively.

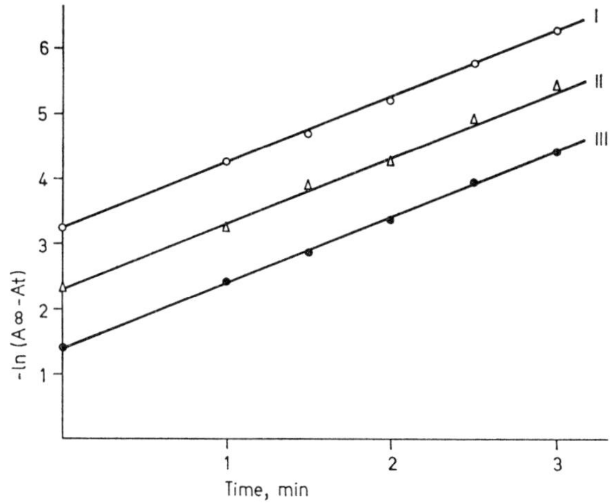

Fig. 1. Time course of reaction for the kinetic assay of uric acid. A_t, absorbance at time t; A_{∞}, absorbance after completion of reaction. The response curves were recorded at Hg 334 nm through an Eppendorf photometer by simulating the GEMSAEC assay. Samples: standards (I, III) or human serum (II).

Results and Discussion

Figure 1 illustrates that the rate of the overall reaction of the assay was pseudo-first order in uric acid concentration.

The linearity of the methods was assayed by measuring appropriate dilutions of a human serum, the uric acid concentration of which had been increased to 1.3 mmol/l by the addition of a weighed quantity of pure uric acid. For the GEMSAEC procedure, we observed an upper limit of linearity of 1.3 mmol/l uric acid (fig. 2). The linearity of the ABA-100 and the LKB 8600 procedure extended up to 1.2 and 0.9 mmol/l uric acid, respectively. As shown in figure 2, the sensitivity of the kinetic method at normal uric acid concentrations was sufficiently high.

Table I summarizes data on the within-run precision of the automated methods. The day-to-day precision was estimated over 10 days using separate aliquots of the same human serum (uric acid content: 384 μmol/l). The coefficient of variation was 2.4% by the ABA-100 method, 1.7% by the GEMSAEC method, and 2.5% by the LKB 8600 method. These results demonstrate the good reproducibility of the kinetic assays.

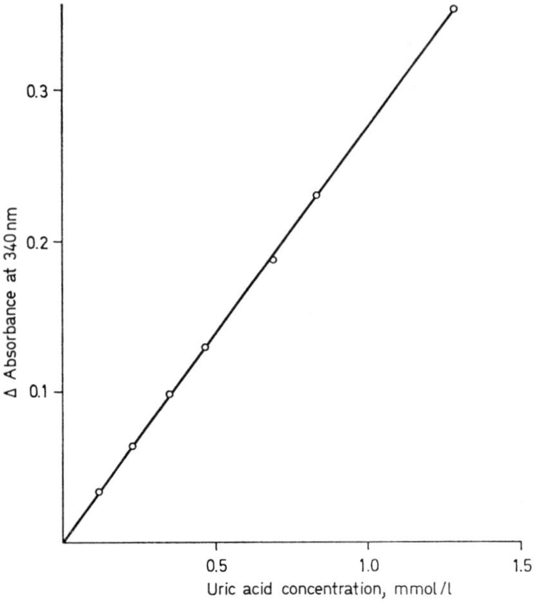

Fig. 2. Proportionality between fixed-time absorbance change and uric acid concentration with the GEMSAEC.

Table I. Within-run precision of kinetic methods for uric acid

Method	Sample	N	Uric acid, mmol/l		CV
			$\bar{x}$	s	%
GEMSAEC	standard	14	0.594	0.003	0.56
	serum	14	0.129	0.002	1.26
	serum	14	0.382	0.002	0.70
	serum	14	1.054	0.010	0.96
ABA-100	standard	20	0.594	0.004	0.78
	serum	20	0.124	0.003	2.67
	serum	20	0.400	0.006	1.46
	serum	20	1.192	0.014	1.18
LKB 8600	standard	20	0.588	0.009	1.59
	serum	20	0.143	0.002	1.92
	serum	20	0.371	0.009	2.58
	serum	20	0.980	0.022	2.22

Table II. Comparison of present automated methods (y) for uric acid with reference methods (x) according to PRAETORIUS *et al.* [2] and to KAGEYAMA [2]

Method	Samples	Data pairs n	Concentration range mmol/l	Regression equation y = bx + a	Correlation coefficient r
PRAETORIUS GEMSAEC	human sera	50	0.148–1.298	y = 0.945x + 0.018	0.997
PRAETORIUS ABA-100	human sera	50	0.148–1.200	y = 0.987x − 0.002	0.998
KAGEYAMA GEMSAEC	human sera	60	0.148–1.298	y = 0.977x + 0.016	0.999
KAGEYAMA ABA-100	human sera	60	0.148–1.200	y = 0.970x + 0.012	0.999
KAGEYAMA LKB 8600	human sera	50	0.148–0.891	y = 0.951x + 0.021	0.988

Analytical recovery for the new procedures was assessed by use of a serum pool to which various amounts of pure uric acid were added. All three methods yielded recoveries of better than 98%.

In comparison studies, results by the kinetic automated methods were compared with results by the UV method according to PRAETORIUS *et al.* [2] and by the colorimetric method according to KAGEYAMA [2]. As shown in table II, there was a satisfactory agreement between the kinetic and the reference methods.

There were no interferences with the kinetic methods from lipemia, hemolysis (up to 42 μmol of hemoglobin/l), bilirubin (up to 171 μmol/l), and various drugs which are known to be inhibitors of aldehyde dehydrogenase [1].

Summary

A kinetic fixed-time method for the determination of serum uric acid was developed using uricase, catalase and aldehyde dehydrogenase. Through the addition of the competitive inhibitor xanthine, the Michaelis constant of uricase was apparently increased, which considerably extended the linearity of the assay. The method has been adapted to the Abbott ABA-100 analyzer, the ENI GEMSAEC centrifugal fast analyzer, and the LKB 8600 reaction rate analyzer. The procedures yielded satisfactory results with regard to precision, accuracy and insensitivity to interferences.

References

1 HAECKEL, R.: The use of aldehyde dehydrogenase to determine H_2O_2-producing reactions. J. clin. Chem. clin. Biochem. *14:* 101–107 (1976).

2 SCHEIBE, P.; BERNT, E., and BERGMEYER, H. U.: Uric acid; in BERGMEYER Methods of enzymatic analysis, vol. 4, pp. 1951–1958 (Academic Press, New York 1974).

3 ZIEGENHORN, J.: Bestimmung der Substanz-Konzentration von Metaboliten (kinetische Methoden); Reaktionskinetische Basis; in BERGMEYER Grundlagen der enzymatischen Analyse, pp. 81–83 (Verlag Chemie, Weinheim 1977).

J. ZIEGENHORN, PhD, Boehringer Mannheim GmbH, Biochemica Werk Tutzing, Forschungszentrum, POB 120, *D-8132 Tutzing* (FRG)

Enzymes in Health and Disease. Inaug. Scient. Meet. Int. Soc. Clin. Enzymol., London 1977, pp. 137–139 (Karger, Basel 1978)

Butanol Extraction of Serum and Urinary γ-Glutamyltransferase

PETER R. BECK
Department of Biochemistry, Royal Infirmary, Glasgow

Butanol was first used in 1950 for the release of enzymes from lipid-containing membrane material [3]. However, most subsequent studies have used butanol as a preparative stage in enzyme purification. γ-Glutamyltransferase (γ-GT, EC 2.3.2.2) is a membrane-bound enzyme, and since it may exist in the circulation attached to membrane fragments [2] it was thought worthwhile to investigate the effects of *n*-butanol on serum γ-GT in a variety of diseases, and to compare this with the enzyme present in urine.

Materials and Methods

γ-GT was assayed kinetically at 37 °C using γ-glutamyl-*p*-nitroanilide as substrate [1]. Serum samples were routine samples received by the laboratory for 'liver function tests', stored at 4 °C until assay. Urines were aliquots from 4-hour collections, also stored at 4 °C.

The method used for extraction of sera and urines consisted of dilution of the sample ×2 with water (this to lessen the gross precipitation of proteins in serum which occurred using neat sample), followed by addition of an equal volume of *n*-butanol. The samples were mixed on a rotary mixer for 15 min followed by centrifugation at 1,400 *g* for 5 min. The lower aqueous layer was used for γ-GT assay within 15 min. The γ-GT activity following butanol extraction was expressed as a percentage of the γ-GT activity (diluted ×2) prior to extraction.

Results

32 sera and 29 urines obtained from healthy laboratory workers were extracted with butanol. Very little γ-GT activity was lost from normal sera; all except two of the sera had more than 75% of activity remaining (fig. 1) and this was taken as the lower limit of normal. On the other hand, no urine had

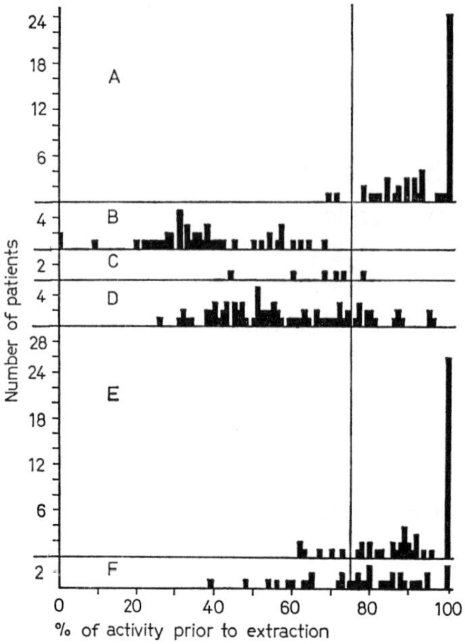

Fig. 1. Results of butanol extraction of serum and urinary γ-GT. Activity following extraction is expressed as a percentage of activity prior to extraction. Vertical axis shows the number of patients in each group. The vertical line at 75% remaining activity is the lower limit of normal for sera. The patient groups are: (A) normal sera; (B) normal urines; (C) hepatobiliary disease (normal total serum γ-GT); (D) hepatobiliary disease (increased total serum γ-GT); (E) other diseases (normal total serum γ-GT); (F) other diseases (increased total serum γ-GT).

more than 68% of activity remaining. Visking tubing dialysis of urines had no effect on butanol extraction. Addition of bovine serum albumin stabilised the urines towards butanol extraction, but only slightly. Activity remaining after butanol extraction of three urines averaged 42%. After addition of 40 g/l albumin, activity averaged 62%.

182 sera from patients with a wide variety of disorders were extracted with butanol. Of 82 patients with increased serum γ-GTs as a result of known hepatobiliary disease 81% had an increased loss of activity following butanol extraction (fig. 1). A further 6 patients with hepatobiliary disease had normal serum γ-GTs. Of these, 5 had low percentage remaining activity after butanol extraction.

The remaining patients suffered from a wide variety of disorders, but in which there was no evidence, biochemically or clinically, of liver involvement.

This included patients with uncomplicated myocardial infarctions. 56 patients had normal serum γ-GTs and 88% of these had normal butanol extractions. Of the 38 patients with increased γ-GTs, only 34% had increased loss of activity. The difference in butanol extraction of increased serum γ-GTs between patients with hepatobiliary disease and those with other disorders was highly significant ($p<0.001$).

Discussion

The differences in loss of activity after butanol extraction of serum and urine correlate well with the Sephadex G-200 gel filtration patterns of γ-GT. Thus, a large peak of γ-GT activity eluted with the void volume is characteristic of urines, and of sera from patients with hepatobiliary disease; i.e. where there is a marked loss of γ-GT activity after butanol extraction. This peak is specifically destroyed following butanol extraction [BECK, in preparation].

There has been considerable discussion [4] as to whether *all* increases of serum γ-GT arise from the liver, even if this is not clinically apparent. The results presented here show that it is possible to distinguish between hepatobiliary and other causes of an increased serum γ-GT activity using butanol, and this may, therefore, suggest different tissue sources for the enzyme in each case. However, an alternative explanation is that the γ-GT in both cases does arise from the liver, but that the differing nature of the insult results in release of different molecular forms of γ-GT.

References

1 BECK, P.R. and CHAUDHURI, A.K.R.: Effect of tobramycin on urinary γ-glutamyltransferase activity: studies in a case of renal carcinoma. Clin. Chem. *22:* 528–531 (1976).

2 BROE, M.E. DE; BORGERS, M., and WIEME, R.J.: The separation and characterization of liver plasma membrane fragments circulating in the blood of patients with cholestasis. Clin. chim. Acta *59:* 369–372 (1975).

3 MORTON, R.K.: Separation and purification of enzymes associated with insoluble particles. Nature, Lond. *166:* 1092–1095 (1950).

4 ROSALKI, S.B.: Gamma-glutamyl transpeptidase; in BODANSKY and LATNER Advances in clinical chemistry, vol. 17, pp. 53–107 (Academic Press, New York 1975).

P.R. BECK, MB, Department of Biochemistry, Royal Infirmary, *Glasgow G4 0SF* (Scotland)

Enzymes in Health and Disease. Inaug. Scient. Meet. Int. Soc. Clin. Enzymol., London 1977, pp. 140–143 (Karger, Basel 1978)

The Nature and Incidence of Particulate Alkaline Phosphatase and Lipoprotein-X in Liver Disease

D. Brocklehurst, C. E. Wilde and J. W. H. Doar

Area Department of Clinical Chemistry, Doncaster Royal Infirmary, Doncaster

Introduction

Particulate alkaline phosphatase (ALP) activity and lipoprotein-X (LP-X) are often found in serum of patients with liver disease [2, 8] particularly when cholestasis is involved. The particulate enzyme is stationary upon starch gel electrophoresis and excluded on Sepharose 4B filtration together with other plasma membrane enzymes. This enzyme complex has similar characteristics to LP-X in the ultracentrifuge and on agar electrophoresis. After fractionation, its particulate nature can be demonstrated by electron microscopy which shows two types of particle. The smaller particle, characteristic of LP-X can be shown to be associated with the enzyme activity by histochemical studies [2].

The purpose of this communication is to assess the relative usefulness of serum particulate ALP and LP-X assays as screening tests of abnormal liver function and to record their incidence in patients with obstructive and non-obstructive liver disease.

Material and Methods

The biochemical indices of liver dysfunction used in the study were total bilirubin, albumin, immunoglobulins A, G and M, ALP, 5-nucleotidase, γ-glutamyl transferase, 1-aspartate and 1-alanine aminotransferase as described in Brocklehurst [1]. The starch gel electrophoresis pattern for ALP isoenzyme was interpreted as abnormal if liver, bone or intestinal bands were present in increased amounts, or if the particulate enzyme, or any tumour-specific bands were present. Rapidophore electrophoresis [6] for LP-X was recorded as positive or negative.

Results

Of 2,841 patients submitted for liver function tests during a 9-month period in a District General Hospital, 715 had a result outside the normal reference range for at least one biochemical test of liver function. The relative sensitivities of these tests in detecting hepatobiliary dysfunction in these patients confirmed that γ-glutamyl transferase was the most sensitive indicator of liver disorder (raised in 87% of patients) and showed that neither the presence of particulate ALP (in 35% of patients) nor that of LP-X (in 23% of patients) was a useful general indicator of hepatobiliary disease.

A definitive diagnosis was obtained in 241 of the 715 patients. These cases were divided into two groups, those with well-defined obstructive liver disease (cholestasis, bile duct stricture, primary biliary cirrhosis, secondary carcinoma of the liver), and those in which significant obstruction was unlikely (cholecystitis without jaundice, congestive cardiac failure, various types of hepatitis, drug jaundice and diseases known to have a secondary effect on the liver).

The majority (70%) of the patients with obstructive liver disease had both particulate ALP and LP-X in their serum, whilst only 15% of those in the other group were positive by both tests. All the sera from patients in the obstructive group were positive for at least one of these tests, whilst in 50% of the patients with other liver disorders neither substance was detected.

The results demonstrate that although the two substances often occur together in the circulation, particularly in patients with obstructive liver disease, they can also appear independently in 35% of the patients studied. The disorders showing the greatest proportion of patients with particulate ALP only were primary and secondary liver tumours and congestive cardiac failure. The majority of patients with LP-X only had alcoholic cirrhosis; these patients had excess serum intestinal ALP but no particulate enzyme. The mean total ALP activity in the patients with only particulate ALP was 440 IU/l in contrast to 240 IU/l in those with only LP-X.

Discussion

Serum particulate ALP activity and LP-X have a close physical association when isolated from the serum of patients with cholestasis [2]. The present study has shown that although the majority of patients with evidence of obstructive liver disease have both substances in their serum, 35% of the patients with various types of liver disease have one or other appearing independently. Particulate ALP appears more frequently than LP-X in patients with liver dysfunction but because either can appear without the other the relative sensitivities of the assay systems are unlikely to account for this.

ALP has been demonstrated at both the canalicular and sinusoidal surfaces of the hepatocyte pericellular membrane, and HAGERSTAND *et al.* [4] related these sites of activity to the slow-moving (particulate) and the fast moving (liver) bands of activity respectively in human serum. Although both types of activity were increased in many liver diseases, the sinusoidal activity was usually the greater. Only in secondary hepatic malignancy were canalicular and sinusoidal activities increased to similar extents. These findings are consistent with the observations made in serum by JENNINGS *et al.* [5].

There is mounting evidence that both these enzymes are synthesised intracellularly with the canalicular enzyme having a lipid moiety attached during its excretion into the circulation. The particulate enzyme in serum and in bile has been shown to take on the characteristics of the sinusoidal enzyme following butan-1-ol extraction [5, 7]. Recent work with another glycoprotein, 5-nucleotidase, has suggested that once the glycoprotein has been channelled into one of the functional domains of the pericellular membrane it is then possible for redistribution to occur to other regions of the surface by lateral diffusion in the plane of the membrane [3]. A reversal of this mechanism would be one explanation of the presence of canalicular (particulate) enzyme in the circulation in obstructive liver disease. Other mechanisms which could play a part are transhepatocyte regurgitation, regurgitation through the interlobular bile duct or ductule into the lymph vessels in the portal area, or seepage of bile between the tight junctions into the blood sinusoids via the intrahepatic space [3].

These routes of communication between the canalicular and sinusoidal surfaces in obstruction could also be implicated in the formation of LP-X which has recently been shown to be dependent on the refluxing of the bile lipoprotein into the circulation [6]. Characteristic LP-X like particles have only been demonstrated in the space of Disse [9], presumably because intracellularly LP-X becomes a structural lipoprotein in the form of plasma membrane from which it may well originate.

We therefore postulate that when sufficient enzyme induction has taken place at the pericellular hepatocyte membrane and sufficient bile lipoprotein has gained access to the sinusoidal surface of the hepatocyte, the particulate ALP/LP-X complex will be identified in the circulation. This may occur as the result of the two substances arising together from the pericellular membrane or from a later association within the circulation. If insufficient enzyme induction has taken place, or too little bile lipoprotein has gained access to the circulation either LP-X or particulate ALP may appear independently.

Summary

Serum particulate alkaline phosphatase and lipoprotein-X assays have been evaluated as tests of liver dysfunction in 715 patients with at least one abnormal biochemical index of liver function. The relative incidence of each of these markers has been determined in the serum of 241 of these patients with well defined liver disease. The findings are discussed in relation to recent discoveries about the pericellular hepatocyte membrane and it is suggested that both these substances originate from this membrane in liver disease.

References

1 Brocklehurst, D.: The nature of serum alkaline phosphatase which is stationary upon starch gel electrophoresis; MPhil, Leeds (1976).

2 Brocklehurst, D.; Lathe, G.H., and Aparicio, S.R.: Serum alkaline phosphatase, 5-nucleotidase and lipoprotein-X in cholestasis. Clin. chim. Acta *67:* 269–279 (1976).

3 Evans, W.H.: Resolving biochemically the hepatocytes multifunctional pericellular membrane. Trends biochem. Sci. *2:* 169–171 (1977).

4 Hagerstrand, I.; Lindholm, K., and Lindroth, Y.: Endothelial and bile canalicular alkaline phosphatase in human liver and serum. Scand. J. clin. Lab. Invest. *36:* 131–135 (1976).

5 Jennings, R.C.; Brocklehurst, D., and Hirst, M.: A comparative study of alkaline phosphatase enzymes using starch gel electrophoresis and Sephadex gel filtration with special reference to high molecular weight enzymes. Clin. chim. Acta *30:* 509–517 (1970).

6 Manzato, E.; Fellin, R.; Baggio, G.; Walch, S.; Neubeck, W., and Seidel, D.: Formation of lipoprotein-X: its relationship to bile compounds. J. clin. Invest. *57:* 1248–1260 (1976).

7 Price, C.P. and Sammons, H.G.: An interpretation of the serum alkaline phosphatase isoenzyme patterns in patients with obstructive liver disease. J. clin. Path. *29:* 976–980 (1976).

8 Seidel, D.; Gretz, H., and Ruppert, C.: Significance of the LP-X test in differential diagnosis of jaundice. Clin. Chem. *19:* 86–91 (1973).

9 Stein, O.; Alkan, M., and Stein, Y.: Obstructive jaundice lipoprotein particles studied in ultrathin sections of livers of bile duct-ligated mice. Lab. Invest. *29:* 166–172 (1973).

D. Brocklehurst, MPhil, Doncaster Royal Infirmary, Area Department of Clinical Chemistry, *Doncaster DN2 5LT* (England)

Enzymes in Health and Disease. Inaug. Scient. Meet. Int. Soc. Clin. Enzymol., London 1977, pp. 144–148 (Karger, Basel 1978)

Paradoxical Pattern of Enzyme Activities in Intrahepatic Cholestasis

D. M. Goldberg, G. Ellis and R. J. Spooner[1]

Hospital for Sick Children, Department of Biochemistry, Toronto, Ont.

Introduction

Intrahepatic cholestasis is a syndrome with many causes. It may occur as a component of several diseases primarily involving the liver and biliary tree. Pregnancy is a common precipitating factor in females with a cholestatic tendency. Administration of oral contraceptives to susceptible females may also precipitate intrahepatic cholestasis (IHC). Many drugs other than oral contraceptives have been reported to trigger IHC, chlorpromazine being one of the best recognized agents in this regard. The list grows larger each passing year. Recognition of the syndrome and its distinction from extrahepatic obstruction is probably the most difficult task in the differential diagnosis of hepatobiliary diseases. During an extensive study of laboratory tests performed on over 1,000 patients suspected of possible disease involving the liver or biliary tract, we had an opportunity to identify a pattern of results commonly present among patients with IHC and infrequently in subjects with other conditions with which the former group might be confused.

Patients and Methods

Intrahepatic cholestasis. 29 cases with this syndrome were diagnosed. In 12, chlorpromazine was the etiologic agent. In 5, IHC was attributed to oral contraceptives. Three

[1] We thank Dr. A. M. Ward who performed the immunological tests not cited in this report and who assisted in the classification of the case material.

cases had cholestatic jaundice of pregnancy, one being for the second time. Two subjects with IHC of undefined etiology had the diagnosis made at exploratory laparotomy. A further 3 subjects were diagnosed as having IHC due to various other causes. Data from these 25 cases only are presented since the other 4 diagnosed as IHC were referred to us when the episode was close to resolution.

Other hepatobiliary diseases studied. (a) Primary biliary cirrhosis (PBC): 24 subjects had typical clinical features with chronic icterus, hepatomegaly and pruritis. (b) Acute viral hepatitis: the case material was divided into 18 with early infectious hepatitis (EIH) and 41 with late infectious hepatitis (LIH). (c) Gallstones: 35 cases with jaundice due to cholelithiasis had the diagnosis confirmed by radiological examination or by laparotomy. (d) Carcinoma of bile duct (CBD): 22 cases had this diagnosis established at laparotomy or autopsy. (e) Carcinoma of pancreas (CAP): 36 cases had jaundice due to pancreatic cancer, the diagnosis being confirmed surgically.

Biochemical procedures. From a large body of tests performed on all subjects, attention is focused in this report upon the following:

Serum total bilirubin concentration by the method of JENDRASSIK and GROF using the Technicon SMA 12–60.

Serum alkaline phosphatase (AP) activity with disodium phenyl phosphate as substrate, using the Technicon SMA 12–60, results being reported as King Armstrong units (KAU) per deciliter at 37 °C.

Serum aspartate and alanine aminotransferase (As-T and Al-T) activities; isocitrate dehydrogenase (ICDH) activity; and 5′-nucleotidase (5NT) activity; all with the LKB Reaction Rate Analyser (LKB Produkter, Bromma, Sweden) at 37 °C, results being expressed as international units (U) per liter.

Results and Discussion

Scrutiny of data from patients with IHC revealed that relative to the serum bilirubin concentration, the enzymes associated with parenchymal liver damage (As-T, Al-T and ICDH) were only modestly elevated, and never reached the activities encountered in acute and chronic hepatitis when that degree of icterus was manifest. In like fashion, those enzymes classically regarded as markers of biliary tract disease (AP and 5NT) rarely reached the activities relative to the serum bilirubin concentration that one would expect in lesions of the extrahepatic biliary tree. To convey the overall impression with which we were struck, data from cases with IHC due to chlorpromazine are assembled in table I, and data from cases with IHC due to oral contraceptives or pregnancy are combined in table II, since the features of these two cholestatic processes suggest that they are different parts of the spectrum of the same clinico-pathological entity.

We attempted to conceptualise this pattern in the form of a number of ratios. Dividing serum AP activity by the bilirubin concentration was not

Table I. Patients with cholestatic jaundice due to chlorpromazine (normal ranges in parentheses)

Case No.	Bilirubin (0–1.0 mg%)	As-T (0–30 U/l)	Al-T (0–30 U/l)	ICDH (0–7 U/l)	5NT (0–11 U/l)	AP (3–15 KAU/100 ml)
1[1]	14.7	35	42	5	14	51
2[1]	29.5	34	90	6	13	24
3[1]	31.4	86	151	9	18	50
4[1]	28.2	145	146	28	16	34
5[1]	5.7	49	87	6	5	24
6[1]	14.0	43	19	7	17	54
7[2]	6.5	54	86	12	11	18
8[2]	11.1	29	39	5	17	29
9	9.9	80	14	2	2	25
10	14.5	30	42	5	31	37
11	12.3	36	26	8	26	27
12	19.5	61	97	4	19	24

[1] Confirmed at biopsy.
[2] Confirmed by reexposure.

Table II. Patients with cholestatic jaundice due to the contraceptive pill (CP) or pregnancy (PR) (normal ranges in parentheses)

Case No.	Bilirubin (0–1.0 mg%)	As-T (0–30 U/l)	Al-T (0–30 U/l)	ICDH (0–7 U/l)	5NT (0–11 U/l)	AP (3–15 KAU/100 ml)
1 CP[1]	8.9	27	35	6	20	21
2 CP[1]	3.6	17	20	4	9	16
3 CP	13.4	49	46	8	6	25
4 CP	18.8	61	73	13	22	19
5 CP	15.7	53	66	4	10	24
6 PR[2]	4.0	36	19	3	50	38
7 PR	7.6	40	36	7	23	22
8 PR	8.1	45	41	4	17	10

[1] Confirmed by reexposure.
[2] Jaundiced during previous pregnancy.

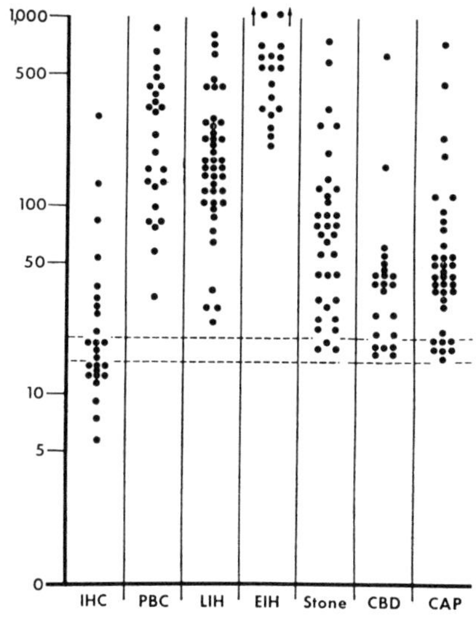

Fig. 1. Values for the quotient AP + 5NT + As-T + Al-T + ICDH divided by serum bilirubin concentration in various disease states. Abbreviations defined in text. Horizontal lines define the limits 15 and 20 on the log scale.

helpful in discriminating IHC from the other disease states. Adding 5NT to the numerator did not greatly improve discrimination. When As-T was added to the numerator, the discrimination was greatly improved. However, best resolution was obtained when the sum of the three previous enzymes together with ICDH and Al-T was divided by the bilirubin concentration. As shown in figure 1, 45% of our IHC cases and no others fell below the value of 15 for the quotient. A further 20% lay between 15 and 20, as did 8% of the remaining case material, or 15% of the cases with extrahepatic disease since all these cases fell into this category.

Obviously, we cannot consider enzyme tests in isolation. Immunological tests proved very helpful in separating IHC from primary liver disease. These tests were also more frequently abnormal in extrahepatic disease, presumably because infection and inflammation are commoner accompaniments of these disease states than in IHC. A clear mechanistic picture emerges for IHC which emphasizes the failure of excretory function with minimal disturbance of the biliary epithelium, minimal disturbance of hepatocellular integrity, and minimal inflammatory response. The pattern we have described is thus helpful in understanding the mechanism of IHC and facilitating its diagnosis.

Summary

A study involving over 1,000 patients with suspected disease of the hepatobiliary system revealed 29 in whom the final diagnosis was intrahepatic cholestasis. Disproportionate elevation of the serum bilirubin concentration relative to serum enzyme activities helped to distinguish intrahepatic from extrahepatic cholestasis. The results emphasize a pure excretory defect as the essential lesion in this condition, with integrity of the biliary epithelium and the hepatocyte being minimally compromised.

D. M. GOLDBERG, MD, Hospital for Sick Children, Department of Biochemistry, 555 University Avenue, *Toronto, Ontario M5G 1X8* (Canada)

Enzymes in Health and Disease. Inaug. Scient. Meet. Int. Soc. Clin. Enzymol., London 1977, pp. 149–153 (Karger, Basel 1978)

The Isoenzymes of Creatine Kinase in Various Diseases

Aarne Konttinen

Kivelä Municipal Hospital, Helsinki

The ultimate criteria for a diagnostic test are its specificity and sensitivity for a diseased organ. Established serum enzyme tests leave a great deal to be desired in this respect and the overwhelming majority of serum enzyme determinations have gained acceptance because of the absence of more accurate tests. Several isoenzymes have facilitated diagnostic accuracy, particularly those of creatine kinase (CK) which are mainly found in skeletal muscle, brain and the myocardium, and have quite different chemical structures. This allows their separation by techniques in common use in clinical laboratories.

The muscle-type CK molecule is composed of two identical subunits (MM). Brain CK is a dimer containing two B subunits, while the myocardium-specific isoenzyme contains both subunits (MB). It has been observed that isoenzymes may be divided into several subfractions [11], though these have not until now had any clinical significance, and are not distinguishable by the methods used for clinical purposes. The only exception is the separation of mitochondrial isoenzymes, where a distinct fraction showing cationic behaviour becomes visible [10].

The diagnostic usefulness of CK isoenzymes is exceptional and superior to the other documented serum tests for the diagnosis of acute myocardial infarction. Although preliminary results for the detection of infarcts were not very convincing due to methodological insensitivity [2], subsequent studies have established its superiority for this purpose (fig. 1). Since negligible activity of the MB isoenzyme occurs outside the myocardium (table I), this offers in clinical practice a specific and most sensitive indicator of myocardial damage (table II). In addition, CK-MB offers the only reliable indicator of myocardial infarction in complicated cases [3].

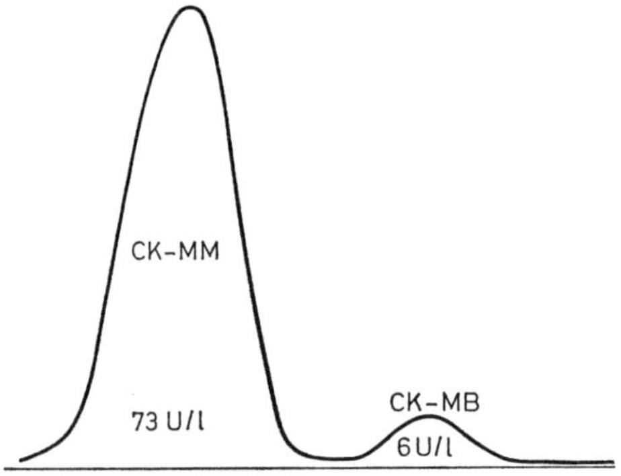

Fig. 1. The scanning of CK isoenzymes in a patient with acute subendocardial infarct. Total serum CK activity 79 U/l, the normal upper limit being 50 U/l.

Surgical procedures, cardiac catheterization and coronary angiography, not uncommonly accompanied by chest pain, arouses suspicion of myocardial infarct. In these situations, heart-type MB isoenzyme is usually of distinctive diagnostic value. Neither congestive heart failure provoked by an infarct or due to heart insufficiency from other reasons, nor haemolysis, disturb the clinical use of MB. The high degree of specificity renders the use of MB invaluable for diagnosing myocardial damage in patients with chest pain and equivocal electrocardiographic changes. A disadvantage of serum CK-MB isoenzyme is its rapid disappearance from the circulation.

Differentiating pulmonary embolism from myocardial infarction is a particularly demanding challenge. This is due largely to the similarities in the clinical symptoms and signs and in the electrocardiographic changes. Earlier results concerning the value of conventional serum enzyme analyses in differentiating between these clinical conditions were conflicting. A comprehensive enzyme study in our laboratories involving numerous isoenzyme analyses, combined with radiological and other techniques, revealed that most of the enzyme activity released into the circulation after pulmonary embolism is derived either from the liver or from haemolysis [4]. The CK isoenzyme elevation is derived entirely from skeletal muscle (MM) since none of the patients had heart-type isoenzyme in their sera. Unfortunately, the BB isoenzyme, characteristic of lung tissue, was not augmented in any of the patients, rendering its use invalid for the confirmation of lung infarction and necessitating repeated radiological study in order to finally establish pulmonary embolism. Other areas (groups of diseases) in which CK iso-

Table I. CK isoenzyme distribution in human tissues which have significance in diagnostic sense

Tissue	CK activity, U/g wet tissue	MM	MB	BB
Muscle	2,100	100	0	0
Heart	390	78	22	0
Brain	150	0	0	100
Lung	9	15	0	85
Kidney	17	10	0	90
Liver	3	93	3	4

Table II. Elevated serum enzyme activities in some diseases which frequently disturb the differentiation from myocardial infarction

	Myocardial infarction	Liver disease	Pulmonary embolism	Muscle disease	Haemolysis
ASAT	4	4	1	2	1
LD	3	2	2	1	2
LD_1	3	0	1	0	3
CK	5	0	0	6	0
CK-MB	6	0	0	0	0

The numbers represent the extent of serum enzyme elevations.

Table III. The prognosis of patients with cerebral accidents having superimposed myocardial injury as indicated by presence of heart-type (MB) isoenzyme

	MB+	MB−
Brain infarction	4/ 6	3/21
Subarachnoid haemorrhage	2/ 3	4/ 5
Cerebral haemorrhage	3/ 3	0/ 1
Cerebral contusion	3/ 9	0/ 1
Bacterial meningitis	2/ 2	0/ 2
Total	14/23	7/30

Of the patients who had MB in their sera, 14 (61%) out of 23 died, whereas of those who did not have MB in their sera, 7 (23%) out of 30 succumbed ($p<0.01$).

enzymes have diagnostic significance are muscle diseases and disorders. The muscle-type isoenzyme determination surpasses other serum enzyme tests in sensitivity and may be helpful in differentiating various muscular diseases. Increased MM isoenzyme in the serum signifies a disorder of skeletal muscle. Normal serum CK activity is derived from the skeletal muscle and is composed entirely of the MM isoenzyme, as determined with the methods in current clinical use. The MM isoenzyme increases in muscle trauma, strenuous exercise, muscular injections, and electric cardioversion.

In myopathies, CK isoenzymes may have value in differentiating various muscular disorders of different aetiologies. In muscular dystrophies, such as Duchenne's, Becker's or congenital dystrophy, the normal muscle-type isoenzyme is partly replaced in the tissue by the MB isoenzyme [8]. This explains why, in these diseases, MB isoenzyme is found in the serum [9]. It can also be of help in diagnosing Duchenne's disease in a child who has not yet developed the typical picture of this hereditary condition. Urgent diagnostic help is needed to detect the female carrier and thus prevent pregnancy and birth of a sick boy. The MB isoenzyme can only be found in a few of these carriers, although total serum CK activity is increased in 3 out of 4 cases [7].

The brain-type isoenzyme of CK was, until recently, thought to be prevented by the blood-brain barrier from escaping into the serum. It has been recently found in the serum, however [5, 11], and the preliminary data suggest that CK-BB may have clinical significance [1]. It is found quite often in diffuse cerebral damage involving large areas of the brain, but BB very seldom appears in the serum in local vascular brain damage. Its diagnostic use is limited by its rapid disappearance from the circulation. The finding of the heart-type isoenzyme in patients with brain accidents having superimposed myocardial damage also has clear prognostic value (table III). An extremely sensitive radioimmunoassay system described recently [6] may be expected to offer additional help in diagnosing brain injuries.

Summary

Serum isoenzymes of creatine kinase are of diagnostic help in the detection of myocardial infarction (MB isoenzyme), brain accidents (BB) and muscle disorders (MM). Myocardial damage can be confirmed conclusively and with great sensitivity by MB determination. In addition, the heart-type isoenzyme is encountered in muscular dystrophies due to partial replacement of the normal MM muscle isoenzyme by the MB-type. The brain-type isoenzyme has been found useful in evaluating the severity of brain damage.

References

1 Kaste, M.; Somer, H., and Konttinen, A.: Brain-type creatine kinase isoenzymes. Archs Neurol., Chicago *34:* 142–144 (1977).

2 Konttinen, A. and Somer, H.: Determination of serum creatine kinase isoenzymes in myocardial infarction. Am. J. Cardiol. *29:* 817–820 (1972).

3 Konttinen, A. and Somer, H.: Specificity of serum creatine kinase isoenzymes in diagnosis of acute myocardial infarction. Br. med. J. *i:* 386–389 (1973).

4 Konttinen, A.; Somer, H., and Auvinen, S.: Serum enzymes and isoenzymes. Extrapulmonary sources in acute pulmonary embolism. Archs intern. Med. *133:* 243–246 (1974).

5 Nealon, D.A. and Henderson, A.R.: Measurement of brain-specific creatine kinase isoenzyme activity in serum. Clin. Chem. *21:* 1663–1666 (1975).

6 Neumeier, D. and Hofstetter, R.: Radioimmunoassay for subunit B in isoenzymes CK-MB and CK-BB of creatine kinase. Clin. chim. Acta *79:* 107–113 (1977).

7 Smith, I. and Thomson, W.H.S.: Carrier detection in X-linked recessive (Duchenne) muscular dystrophy: pyruvate kinase isoenzymes and creatine phosphokinase in serum and blood cells. Clin. chim. Acta *78:* 439–451 (1977).

8 Somer, H.; Donner, M.; Murros, J., and Konttinen, A.: A serum isozyme study in muscular dystrophy. Archs Neurol., Chicago *29:* 343–345 (1973).

9 Somer, H.; Dubowitz, V., and Donner, M.: Creatine kinase isoenzymes in neuromuscular diseases. J. neurol. Sci. *29:* 129–136 (1976).

10 Somer, H.; Uotila, A.; Konttinen, A., and Saris, N.E.: Creatine kinase activity and its isoenzyme pattern in heart mitochondria. Clin. chim. Acta *53:* 369–372 (1974).

11 Wewers, R.A.; Olthuis, H.P.; Niel, J.C.C. van; Wilgenburg, M.G.M. van, and Soons, J.B.J.: A study on the dimeric structure of creatine kinase. Clin. chim. Acta *75:* 377–385 (1977).

A. Konttinen, MD, Kivelä Municipal Hospital, Sibeliuksenkatu 12–14, *SF-00260 Helsinki 26* (Finland)

Enzymes in Health and Disease. Inaug. Scient. Meet. Int. Soc. Clin. Enzymol., London 1977, pp. 154–159 (Karger, Basel 1978)

Determination of Creatine Kinase Isoenzyme MB from Extracardial Sources in Serum[1]

G. CHEMNITZ, F. W. SCHMIDT, E. SCHMIDT and J. LOBERS

Division of Gastroenterology and Hepatology, Department of Internal Medicine, Medizinische Hochschule Hannover, Hannover

Introduction

In myocardial infarction, the ECG parameters are atypical for 20% of the patients. The determination of creatine kinase (CK, EC 2.7.3.2) isoenzyme MB in serum is therefore used more and more for early and differential diagnosis in myocardial infarction. In heart muscle, the proportion of CK-MB is highest with 22% [2]. CK-MB, however, is not only a myocardium-specific isoenzyme but is also present in many other parenchymal tissues [2, 4, 6, 9]. Thus, it is necessary to know whether CK-MB activity in serum can also be detected in extracardial diseases.

Patients and Methods

CK isoenzymes were differentiated in serum of 198 patients with extracardial diseases and increased total CK activity. In addition, we investigated CK-MB activity in 429 samples of patients with myocardial infarction in different phases after the onset of pain. Total CK, GOT and GPT activities were assayed according to the recommendation of the German Society of Clinical Chemistry with the *Boehringer-Mono test* or *Merck-1 test.* CK isoenzymes were differentiated with two immunological methods: immunoinhibition [4, 11] and – only in extracardial cases with measured CK-B subunit activity – immunoprecipitation [1]. Photometer Eppendorf, 334 nm, +25 °C.

[1] Supported by a grant from the Deutsche Forschungsgemeinschaft, Bonn-Bad Godesberg.

Table I. Diagnoses of patients with extracardial CK-MB activity in serum

Diagnosis	CK-MB, U/l		CK-MB, %		n
	x̄	range	x̄	range	
Muscle disease	51	4–552	5	0.3–9	21
Intoxications	45	4–192	6	4–8	18
Shock	31	6–97	5	2–9	8
Various traumata	18	5–92	4	0.7–14	10
Delirium tremens	135	4–796	5	2–10	9
Cerebrovasc. accidents	9	4–20	3	2–5	4
Thyroid diseases	12	4–21	5	0.8–13	14
Psychosis, acute	14	4–32	6	2–8	6
After delivery	11	5–18	5	2–9	4
Glycogenosis type II	5	5	4	4	1
Erysipelas	8	8	5	5	1
Diabetic coma	12	12	7	7	1
Others	18	9–48	7	3–15	5

Results

The diagnoses of patients with extracardial diseases and elevation of CK-MB activity in serum are listed in table I.

102 of 198 patients with extracardial diseases and increased total CK had CK-MB activity in serum of more than 4 U/l. The frequency distribution of the samples shows that 50% of all values had CK-MB activity between 4 and 15 U/l. The rest show a wide distribution of values from 20 to more than 100 U/l. The peak CK-MB activity was 800 U/l, measured in a patient with delirium tremens and total CK activity of 13,000 U/l.

The frequency distribution of the relative amount of CK-MB shows that 50% of all values varies between 4 and 7%. The lowest was 0.3%, the highest 14%.

80% of all patients with myocardial infarction have the typical ECG parameters [5], and no enzyme activity is necessary for differential diagnosis. The determination of CK-MB becomes especially important when signs and symptoms, biochemical or ECG parameters can neither exclude nor confirm acute myocardial infarction. The differentiation of CK-MB activity from extracardial sources from CK-MB activity after myocardial infarction is decisive for clinical purposes, but the CK-MB values of extracardial diseases varied from 4 to 800 U/l and after myocardial infarction from 4 to 330 U/l. Such overlapping makes discrimination impossible.

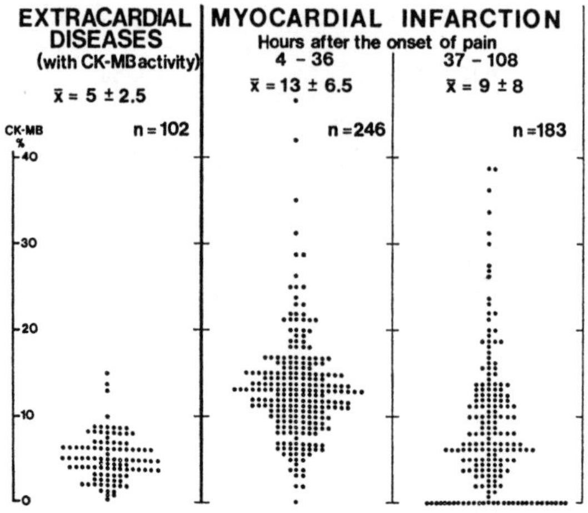

Fig. 1. Relative amount of CK-MB in extracardial diseases and after myocardial infarction.

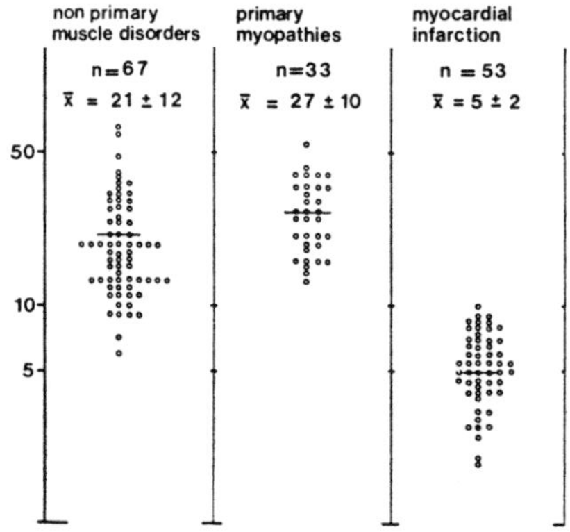

Fig. 2. CK : GOT ratio with a NAC-activated CK test and optimized standard GOT test in myocardial infarction, primary myopathies and non-primary muscle disorders. Criteria: CK activity more than 150 U/l; no liver disease.

The mean of the relative amount of CK-MB from extracardial disorders is 5±2.5%. Between 4 and 36 h after myocardial infarction, the relative amount of CK-MB is 13±6.5% (fig. 1). Figure 1 shows that a CK-MB fraction above 10% in serum is probably due to myocardial infarction.

When CK-MB fraction allows no statement for differential diagnosis, CK:GOT ratio may be useful for further discrimination. The criteria for using the CK : GOT ratio are (a) total CK activities above 150 U/l, and (b) no liver disease. The original SZASZ ratio was established with a GSH-activated CK test and non-optimized GOT test. The discrimination point was formerly 8.9 [7, 8]. We investigated CK:GOT ratios with N-acetyl cysteine-activated CK test and GOT determination under optimized standard condition according to the German Society for Clinical Chemistry. Figure 2 shows that the CK:GOT ratio is a good parameter for discriminating heart and skeletal muscle disease, and that the discriminating point only shifted from 8.9 to 10.

Discussion

Since CK-MB activity is found in serum of many patients with extracardial diseases and increased total CK activity, many diagnostic problems are encountered. The vital factor is how to differentiate between CK-MB activity from extracardial diseases with or without heart muscle involvement and CK-MB activity after myocardial infarction. The difference of the relative amount of CK-MB is statistically significant ($p<0.001$), but the values are derived from a wide range of variation and frequently no statement for the individual patient can be made. When discrimination with CK-MB fraction is not possible, the CK:GOT ratio may be a useful parameter for further differentiation, if proper prerequisites are observed.

The question of the origin of CK-MB activities in extracardial diseases is difficult to answer. It is known that in primary myopathies similar morphological alterations are found in heart muscle as in striated muscle [3]. One may postulate that the increase of CK-MB activity in primary and secondary myopathies may partly be accounted for by release from myocardium. CK-MB activity in serum may be an expression of heart muscle involvement in patients with extracardial diseases.

On the other hand, CK-MB is demonstrated in skeletal muscle. The injured and regenerating muscle contains a higher amount of CK-MB [10]. Nevertheless, in spite of some hypotheses, it frequently turns out to be difficult to explain the origin of CK-MB activities in extracardial diseases.

Conclusions

(1) 102 of 198 patients with extracardial diseases and increased total CK activity had significant CK-MB activity in serum.

(2) A CK-MB fraction above 10% of total CK is probably due to myocardial infarction.

(3) The CK:GOT ratio, measured by NAC-activated CK test and optimized standard GOT test is a good tool for discriminating heart and skeletal muscle disease – if proper conditions are observed.

(4) The results show that taking CK-MB simply as an isoenzyme specific for myocardial infarction may lead to diagnostic errors.

Summary

Since CK-MB is not only a myocardium-specific isoenzyme but is also present in many other parenchymal tissues, it is necessary to know that CK-MB activity in serum can also be detected in extracardial diseases. CK isoenzymes were differentiated quantitatively by immunoprecipitation and immunoinhibition. We succeeded to demonstrate significant CK-MB activity in serum in the following diseases: inflammatory and degenerative muscle diseases, intoxications, delirium tremens, shock, various traumatic lesions, apoplexia, hypothyroidism, acute psychosis and in sera obtained of women after delivery. Serum CK-BB activity was never seen in this series. The application of the CK:GOT ratio for differentiation of skeletal and heart muscle disease with a NAC-activated CK test and optimized standard GOT test is discussed in comparison with CK isoenzyme ratio.

References

1 JOCKERS-WRETOU, E.; GRABERT, K. und PFLEIDERER, G.: Quantitative immunologische Bestimmung der Isoenzyme der Creatinkinase im Serum. Z. klin. Chem. klin. Biochem. *13:* 85–88 (1975).

2 JOCKERS-WRETOU, E. and PFLEIDERER, G.: Quantitation of creatine kinase isoenzymes in human tissues and sera by an immunological method. Clin. chim. Acta *58:* 223–232 (1975).

3 MÖSSLACHER, H.; SLANY, J.; SCHMOLINER, R.; PAMPERL, H.; STOCKINGER, L. und SLUGA, E.: Herzmuskelbeteiligung bei einer Sonderform der Erbschen Muskeldystrophie (Typ Becker-Kiener). Dt. med. Wschr. *102:* 1123–1125 (1977).

4 NEUMEIER, D.; KNEDEL, M.; WÜRZBURG, U.; HENNRICH, N. und LANG, H.: Immunologischer Nachweis von Creatinkinase MB im Serum beim akuten Myokardinfarkt. Klin. Wschr. *53:* 329–333 (1975).

5 PRELLWITZ, W.; NEUMEIER, D.; KNEDEL, M.; LANG, H.; WÜRZBURG, U.; SCHÖNBORN, H. und SCHUSTER, H.P.: Isoenzyme der Kreatinkinase bei extrakardialen Erkrankungen und nach diagnostischen und therapeutischen Eingriffen. Dt. med. Wschr. *101:* 983–988 (1976).

6 ROBERTS, R.; GOWDA, K.S.; LUDBROOK, P.A., and SOBEL, B.E.: Specificity of elevated serum MB creatine phosphokinase activity in the diagnosis of myocardial infarction. Am. J. Cardiol. *36:* 433–437 (1975).

7 Schmidt, E. und Schmidt, F.W.: Optimale Bedingungen zur Bestimmung der Transaminasen und der Kreatin-Phosphokinase im Serum und ihre Konsequenzen für die Diagnostik. Z. analyt. Chem. *243:* 398–414 (1968).

8 Szasz, G. und Busch, E.W.: Organspezifizierung einer erhöhten Kreatinkinase-Aktivität im Serum; in Kaiser Fortschritte der Klinischen Chemie, Enzyme und Hormone, pp. 163–167 (Verlag der Wiener Medizinischen Akademie, Wien 1972).

9 Tsung, S.H.: Creatine kinase isoenzyme pattern in human tissue obtained at surgery. Clin. Chem. *22:* 173–175 (1976).

10 Wiesmann, U.; Kaspar, U., and Mummenthaler, N.: Necrosis and regeneration of tibialis anterior muscle in rabbit. Archs. Neurol., Chicago *21:* 373–380 (1969).

11 Würzburg, U.; Hennrich, N. und Lang, H.: Bestimmung der Aktivität von Creatinkinase MB im Serum unter Verwendung inhibierender Antikörper. Klin. Wschr. *54:* 357–360 (1976).

G. Chemnitz, MD, Medizinische Hochschule Hannover, Division of Gastroenterology and Hepatology, Department of Internal Medicine, Karl-Wiechert-Allee 9, *D-3000 Hannover 61* (FRG)

Enzymes in Health and Disease. Inaug. Scient. Meet. Int. Soc. Clin. Enzymol., London 1977, pp. 160–163 (Karger, Basel 1978)

Errors in Enzymatic Estimation of Myocardial Infarct Size due to Changes in Plasma Volume

MOGENS HØRDER, B. LYAGER NIELSEN, P. HYLTOFT PETERSEN and K. THYGESEN

Departments of Clinical Chemistry and Internal Medicine B, Odense University Hospital, Odense

Introduction

In mathematical models [4] used for calculation of the size of myocardial infarct from the plasma enzyme curve, no corrections are usually made for the effects of plasma volume changes on the results. We have determined the over- and underestimation of infarct sizes when the concentration of α_2-macroglobin in plasma was used as indicator for changes in plasma volume concomitant with the infarct.

Materials and Methods

The investigation includes 21 patients with acute myocardial infarction admitted, on average, 2 h after the onset of symptoms. The course of the disease was complicated with slight heart failure in 17 patients, with pulmonary oedema in 2 patients, and with cardiogenic shock in 1 patient. Renal function and electrolyte balance showed no disturbances.

On admission, an intravenous cannula was inserted in the antecubal vein. 5-ml blood samples were drawn into heparin-containing tubes at 3–4 hourly intervals for the following 2 days and then 3–4 times daily up to day 7. The blood samples were stored at +4 °C until plasma was separated, always within 4 h. The catalytic concentrations of creatine kinase (EC 2.7.3.2, CK), aspartate aminotransferase (EC 2.6.1.1, ASAT) and lactate dehydrogenase (EC 1.1.1.27, LD) were determined according to the Nordic Recommended Methods [1, 2]. α_2-Macroglobulin concentration was determined by electroimmunoassay technique.

The catalytic concentrations of the enzymes were corrected by multiplying with the correction factor $[A_o/A_t]$. A_o is the concentration of α_2-macroglobulin in the first one to two blood samples drawn, and A_t is the concentrations of α_2-macroglobulin determined in the following samples together with the catalytic concentrations of enzymes. From the plasma curves for the uncorrected and corrected enzyme data (catalytic concentration vs time) the corresponding cumulative enzyme release curve and the size of the myocardial infarct were calculated. The mathematical model and computer program described by MATHEY *et al.* [3] were used.

Results

The observed, uncorrected plasma CK curve and the cumulative, uncorrected and corrected plasma CK curves in one of the patients are shown in figure 1. The uncorrected cumulative curve shows a plateau approximately 1,000 U/*l* above the corrected curve. The size of the calculated infarct is therefore also bigger when the uncorrected enzyme data are used for calculation, namely a fractional overestimation of 0.12 ([uncorrected-corrected]/uncorrected).

The range of fractional overestimations and underestimations of infarct size by CK, ASAT and LD in the 21 patients are shown in figure 2. The range is from a fraction of +0.27 of overestimation to a fraction of –0.23 of underestimation. In 12 of 21 patients, infarct sizes were erroneous estimated

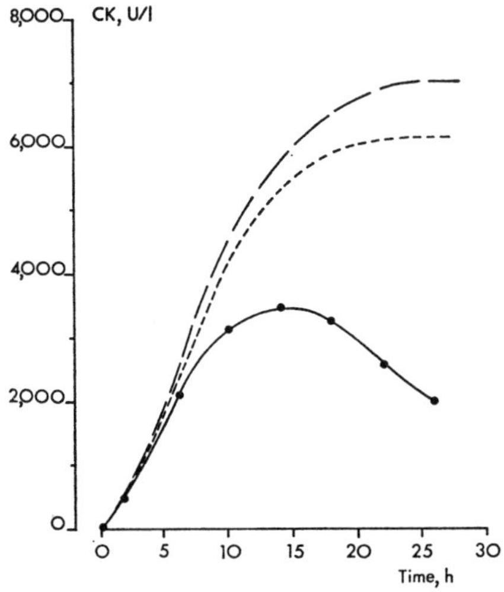

Fig. 1. The observed plasma CK curve (•) in one patient, uncorrected for changes in plasma volume. The cumulative CK release curves, uncorrected and corrected for changes in plasma volume, are the upper (— — —) and middle (– – – –) curves, respectively.

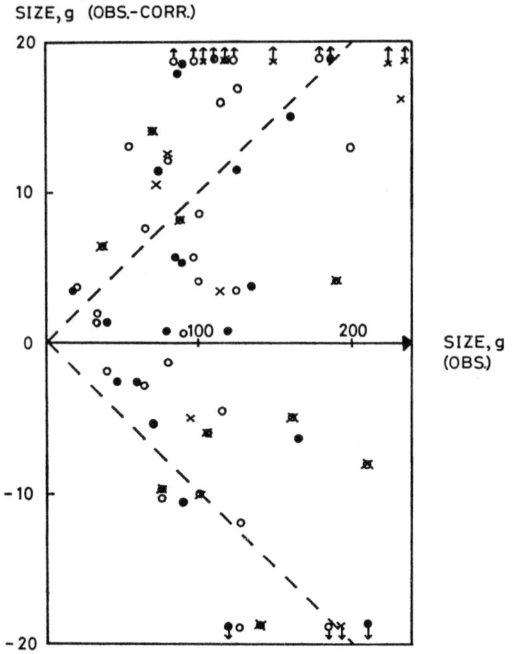

Fig. 2. The effect of correction for changes in plasma volume on the calculated infarct sizes in 21 patients. The abscissa shows the uncorrected size of the infarcts, calculated on basis of the plasma curves of CK (•), ASAT (○) and LD (×). On the ordinate is given the differences between the uncorrected and corrected infarct sizes. The dotted lines indicate a fractional change of +0.1 and –0.1 (upper and lower, respectively) in such a way that fractional changes greater than 0.1 are shown to the left of the dotted lines.

by fractions within +0.10 to –0.10. In 13 patients, the corrections made resulted in a decrease in infarct size (the observed size was overestimated), in 7 patients in an increase, and in only one patient the infarct size did not change by correction.

Discussion

Changes in plasma volume concomitant with myocardial infarct and its influence on quantitation of infarct size has been described by Smith *et al.* [5]. By using hematocrit as indicator of plasma volume changes they observed fractional overestimations of an average 0.12 (range 0.01–0.30). They did not find any sign of underestimation, i.e. increased plasma volume. The basis for the corrections made in this investigation are the assumptions firstly that α_2-macroglobulin does not leak from the plasma compartment during changes in plasma volume, and secondly that its synthesis and catabolism are unchanged in the early period of myocardial infarction. Whether these assumptions are valid remain to be shown.

However, α_2-macroglobulin, due to its molecular size and form, seems preferable as correction factor for enzyme concentrations as compared to the erythrocyte, through the hematocrit.

References

1 Committee on Enzymes of the Scandinavian Society for Clinical Chemistry and Clinical Physiology: Recommended methods for the determination of four enzymes in blood. Scand. J. clin. Lab. Invest. *33:* 291–306 (1974).

2 Committee on Enzymes of the Scandinavian Society for Clinical Chemistry and Clinical Physiology: Recommended method for the determination of creatine kinase in blood. Scand. J. clin. Lab. Invest. *36:* 711–723 (1976).

3 Mathey, D.; Bleifeld, W.; Buss, H., and Hanrath, P.: Creatine kinase release in acute myocardial infarction: correlation with clinical, electrocardiographic, and pathological findings. Br. Heart J. *37:* 1161–1168 (1975).

4 Roe, C.R. and Starmer, C.F.: A sensitivity analysis of enzymatic estimation of infarct size. Circulation *52:* 1–5 (1975).

5 Smith, S.J.; Bos, G.; Hagemeyer, F.; Hermens, W.T., and Witteveen, A.G.J.: Influences of changes in plasma volume on quantitation of infarct size in man by means of plasma enzyme levels. J. mol. Med. *1:* 199–210 (1976).

M. Hørder, MD, Department of Clinical Chemistry, Odense University Hospital, *DK-5000 Odense C* (Denmark)

Enzymes in Health and Disease. Inaug. Scient. Meet. Int. Soc. Clin. Enzymol., London 1977, pp. 164–172 (Karger, Basel 1978)

Differential Diagnostic Value of CK-MB Activity Measurements

D. NEUMEIER, W. PRELLWITZ and M. KNEDEL

Institut für Klinische Chemie, Klinikum Grosshadern, Universität München, München, and Zentrallaboratorium der Medizinischen Kliniken und Poliklinik, Universität Mainz, Mainz

Introduction

We investigated the CK-MB activity in sera of 83 patients with proven myocardial infarction and 229 patients after operations or skeletal muscle traumatization. The CK-MB activity was measured by means of the immunological inhibition method.

We found CK-MB activity in serum in each case of myocardial infarction within 12–24 h after the cardiac event. CK-MB activity was detected, however, in 85% and 92% of all patients in the postoperative period and after polytraumatization, respectively. For this reason, the detection of CK-MB activity in serum only is not in all cases proof of cardiac damage since CK-MB is also freed from damaged skeletal muscle. For a differential diagnostic assessment regarding cardiac muscle damage, the level of percentage CK-MB activity and not the detection of CK-MB activity is the decision criterion. According to our results, this level is in the region of 4–6% CK-MB activity.

In more recent literature, conflicting results concerning the organ distribution of CK isoenzymes have been made. While THORSTENSSON *et al.* [1] give a value of 30% as percentage activity of CK-MB isoenzyme for both human heart and skeletal muscle, SOBEL's team [2], using a chromatographic method, report 15% as percentage activity in heart muscle and none in skeletal muscle. JOCKERS-WRETOU and PFLEIDERER [3] found by means of the immunoprecipitation method a CK-MB activity of 26% and up to 3% in heart muscle and normal skeletal muscle, respectively. During our own investigation of tissue samples using the immunological inhibition method, we found mean values of 2.7 and 42% for CK-MB activity in skeletal and heart muscle, respectively [4, 5]. On account of its high activity in heart muscle, the CK-MB isoenzyme presents itself as the critical enzyme in cases of cardiac damage.

It must be asserted, however, that CK-MB isoenzyme is not in the strictest sense specific to heart muscle tissue, at least when using sensitive immunological methods. How does this affect the clinical differential diagnosis?

Material and Methods

We investigated the activity kinetics of CK total and CK-MB in 83 patients with myocardial infarction (proven by autopsy or clinical and electrophoretic criteria). Serum samples were taken at intervals of 2–6 h. In the same manner, we investigated sera for released CK-MB activity from 229 patients after operations or with skeletal trauma caused by accidents.

The CK-MB activity was measured with the immunological inhibition method and also in some cases by means of the immunological precipitation method, recently published by us [6, 7]. The CK total activity was determined by the method and with the reagents recommended by the German Society for Clinical Chemistry [8]. The upper limit of normal values for CK total activity by this method is 50 U/l.

Results and Discussion

During investigations of the kinetics of CK total and CK-MB activity after myocardial infarction, CK-MB activity could be determined in each case by means of the immunological inhibition method (fig. 1). The ratio of maximum CK total to maximum CK-MB activity is approximately 8 : 1 with a relatively large scatter of the individual activities due to the clinical environment. This ratio is only of limited value in clinical situations since frequently peak activities of CK total and CK-MB do not occur simultaneously relative to infarction [9].

For 31 patients, the elimination constant for CK total and CK-MB could be calculated according to the formula given by SHELL *et al.* [10]. The elimination constant for CK-MB, with a large scatter of the individual values, was with a mean of 9.2×10^{-4} (U/min) on average 25% higher than that for CK total (fig. 2).

For a better understanding of CK-MB activity levels after myocardial infarction, figure 3 shows on a 6-hour time base following the cardiac event

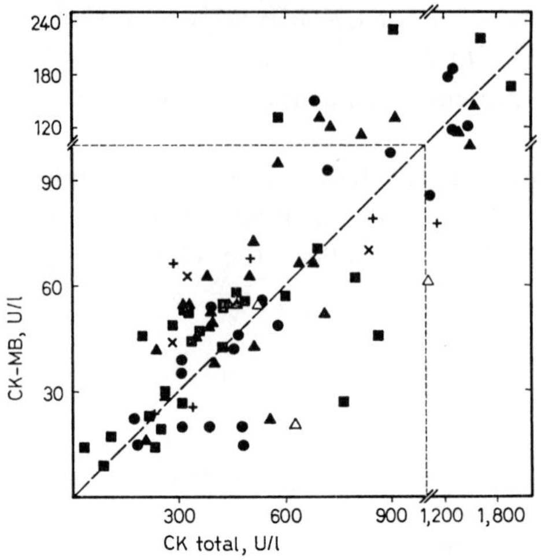

Fig. 1. Maximum CK total and CK-MB activities after myocardial infarction Symbols for the localization of the myocardial infarction: ● = anterior; ▲ = posterior; □ = other; + = not determined; △ = posterior reinfarction; × = not determined reinfarction.

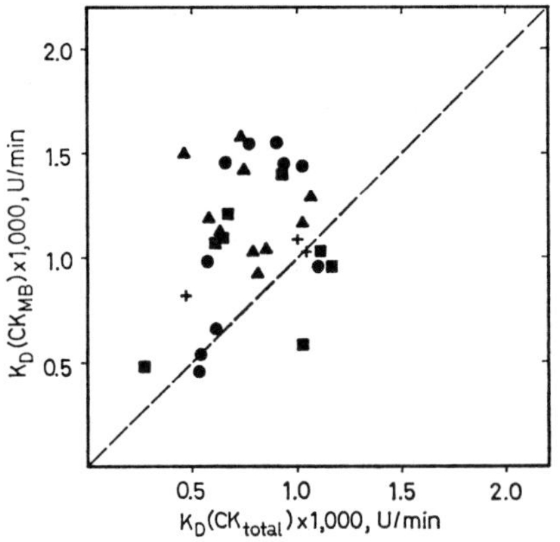

Fig. 2. Elimination constants of CK total and CK-MB activities in serum after myocardial infarction. Symbols for the localization of the myocardial infarction: ● = anterior; ▲ = posterior; ■ = other; + = not determined.

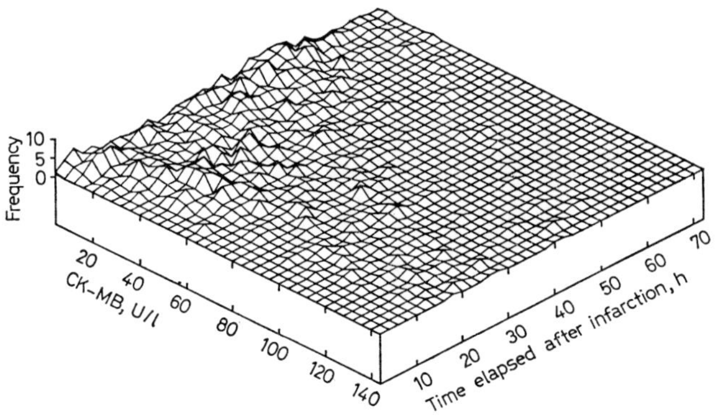

Fig. 3. Frequency distribution of measured CK-MB activities after myocardial infarction.

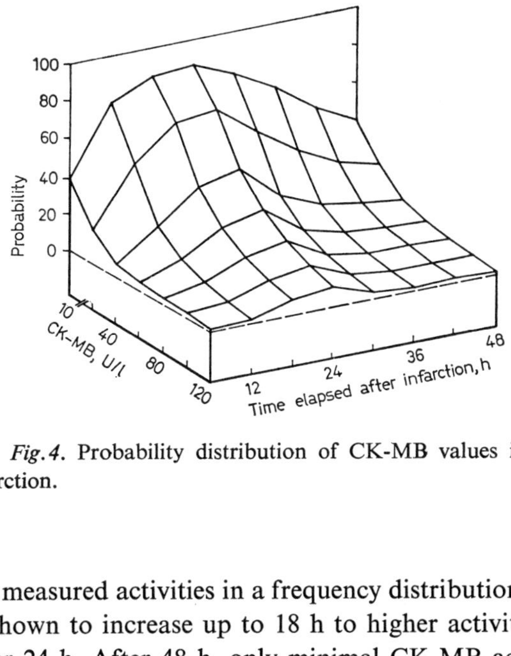

Fig. 4. Probability distribution of CK-MB values in cases of proven myocardial infarction.

the measured activities in a frequency distribution. The maximum frequency is shown to increase up to 18 h to higher activities and to fall away again after 24 h. After 48 h, only minimal CK-MB activities could be measured in the majority of cases due the shorter half-life of isoenzyme CK-MB. In confirmed cases of myocardial infarction, on the other hand, the probability that CK-MB activity is greater than 10 U/l within 24 h after the onset of pain is 92% (fig. 4).

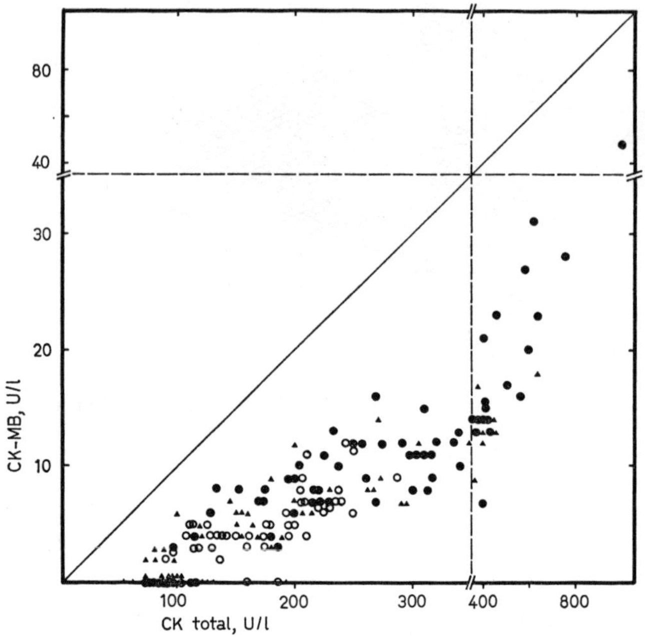

Fig. 5. Maximum CK total and CK-MB activities in postoperative patients. ● = Patients after neurosurgery; ○ = patients after abdominal surgery; △ = patients after orthopedic surgery.

To the clinician, this figure shows that after an uncomplicated myocardial infarction with moderate CK total activities, a CK-MB determination is only of diagnostic significance within 48 h after onset of pain.

The results of the CK-MB determinations in sera of patients in the postoperative phase are shown in figure 5. Isoenzyme CK-MB was present in 85% of all cases. In some cases with extremely high CK total activity, CK-MB activity up to 40 U/l was recorded. High CK-MB activities following neurosurgery were striking. Since the presence of isoenzyme CK-BB has been asserted in the literature [11] following such operations, these sera were at the same time inspected by means of the immunological precipitation method: isoenzyme CK-BB was not, however, found in any of the sera.

Similar results are obtained for patients subjected to polytrauma (fig. 6). In 92% of all cases, isoenzyme CK-MB was found, in some with very high activity. The higher CK-MB activities in cases with thorax involvement

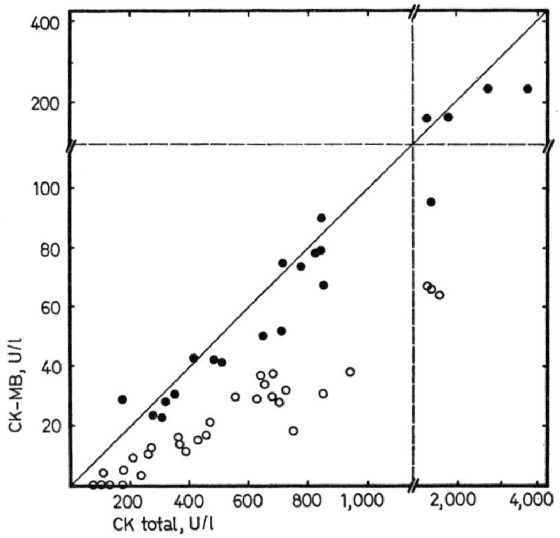

Fig. 6. Maximum CK total and CK-MB activities in patients after polytraumatization: ● = polytraumatization with thorax involvement; ○ = polytraumatization without thorax involvement.

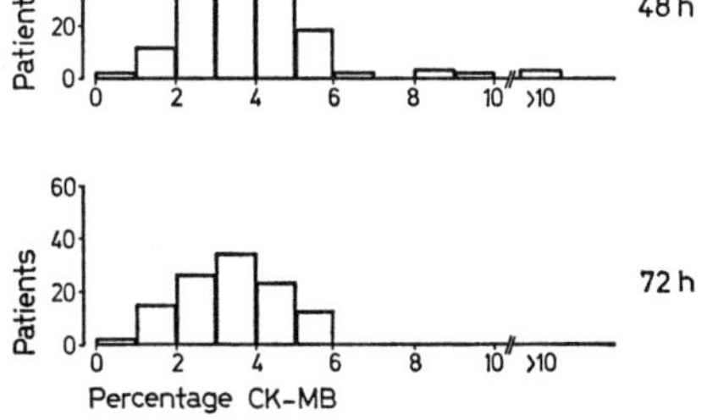

Fig. 7. Frequency distribution of percentage CK-MB activity after skeletal muscle damage.

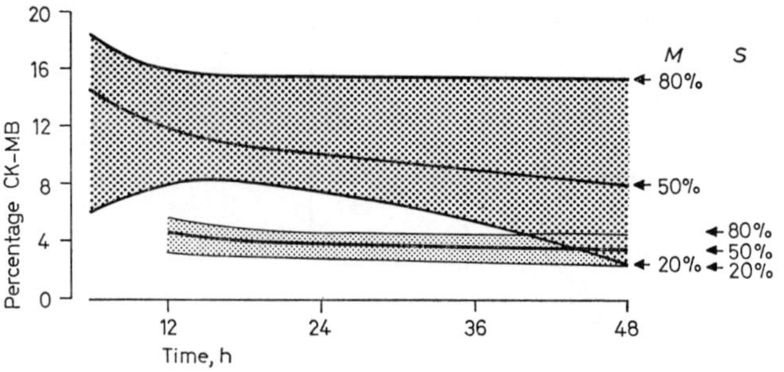

Fig. 8. Percentiles of CK-MB percentage activity after myocardial infarction or traumatization of skeletal muscle. The thick line represents the activity median. The region between the 20 and 80% percentiles is shaded in light grey. Heavy screen and *M* = myocardial infarction; light screen and *S* = traumatization of skeletal muscle.

are noteworthy. Since isoenzyme CK-BB could not be found in any of these sera, the possibility of traumatic damage of the heart has to be discussed in these cases.

Given that CK-MB activity is found in sera after damage of striated muscle, merely the presence of CK-MB activity is insufficient in a clinical assessment that heart damage has taken place. A decision whether heart muscle tissue is damaged or not should rather be made on the isoenzyme pattern in serum or the percentage CK-MB activity of total activity.

The frequency distribution of percentage CK-MB activity after skeletal muscle trauma for cases in which CK-MB activity could be measured has a frequency maximum for a percentage CK-MB activity of 4% independent of sampling time (fig. 7). In a few cases with thoracic involvement, the MB values exceeded 6%.

Figure 8 shows the percentage CK-MB activity levels both after myocardial infarction and skeletal muscle trauma. In the case of myocardial infarction, there is a shift of the frequency maximum to lower percentage activities after 12 h due to the shorter half-life of CK-MB isoenzyme in the serum. After skeletal muscle damage, a time dependence of percentage CK-MB activity is not so pronounced. Only 36–48 h after the cardiac event is there a significant overlapping of the two frequency distributions. The much smaller scatter of percentage CK-MB activity relative to absolute CK-MB activity is important.

According to the difference in percentage CK-MB activity after myocardial infarction and skeletal muscle damage, we recommend the following

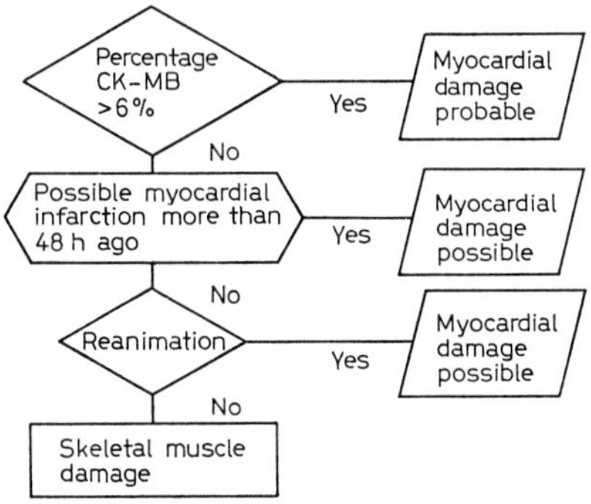

Fig. 9. Decision sequence for patients with proven CK-MB activity in serum.

procedure for the use of CK isoenzyme determination in differential diagnosis (fig. 9). CK-MB activity should only be measured after CK total activity has been determined. If CK-MB activity has been found, a check should be made whether percentage CK-MB activity exceeds 6%. If this is the case, heart muscle damage is probable. If the value is under 6%, it should be ascertained whether the cardiac event in question occurred more than 48 h ago, because of the shorter half-life of CK-MB isoenzyme. If this is not the case either, resuscitation should also be excluded since the additional damage to skeletal muscle tissue might lead to the accumulation of two activity kinetics. As a consequence, the initially high cardiac MB activity would be overlaid by muscle CK-MM activity. If the patient is not resuscitated, it is probable that the CK-MB activity found is due to a release from skeletal muscle.

References

1 Thorstensson, A.; Elwin, K.; Sjödin, B., and Karlsson, J.: Isoenzymes of creatine phosphokinase and myokinase in human heart and skeletal muscle. Scand. J. clin. Lab. Invest. *36:* 831–826 (1976).

2 Roberts, R.; Gowda, K.S.; Ludbrook, P.A., and Sobel, B.E.: Specificity of elevated serum MB creatine phosphokinase activity in the diagnosis of acute myocardial infarction. Am. J. Cardiol. *36:* 433–437 (1977).

3 Jockers-Wretou, E. and Pfleiderer, G.: Quantitation of creatine kinase isoenzyme in human tissues and sera by an immunological method. Clin. chim. Acta *58:* 223–232 (1975).

4 NEUMEIER, D.; KEMKES, B.; GLÜCK, B. und KNEDEL, M.: Verteilung der Creatin-kinase-Isoenzyme in Skelett- und Herzmuskulatur des Menschen. J. clin. Chem. clin. Biochem. *15:* 179–180 (1977).

5 PRELLWITZ, W.; KAPP, S.; NEUMEIER, D.; KNEDEL, M.; LANG, H. und HEUWINKEL, D.: Isoenzyme der Kreatin-Kinase: Verteilungsmuster in der Skelettmuskulatur und im Serum bei Erkrankungen und Schädigungen der Muskulatur. Klin. Wschr. *56:* 559–565 (1978).

6 NEUMEIER, D.; PRELLWITZ, W.; WÜRZBURG, U.; BRUNDOBLER, M.; OLBERMANN, M.; JUST, H.-J.; KNEDEL, M., and LANG, H.: Determination of creatine kinase isoenzyme MB activity in serum using immunological inhibition of creatine kinase M subunit activity. Activity kinetics and diagnostic significance in myocardial infarction. Clin. chim. Acta *73:* 445–451 (1976).

7 WÜRZBURG, U.; HENNRICH, N.; ORTH, H.-D.; LANG, H.; PRELLWITZ, W.; NEUMEIER, D.; KNEDEL, M., and RICK, W.: Quantitative determination of creatine kinase isoenzyme catalytic concentrations in serum using immunological methods. J. clin. Chem. clin. Biochem. *15:* 131–137 (1977).

8 Deutsche Gesellschaft für Klinische Chemie: Standardisierung von Methoden zur Bestimmung von Enzymaktivitäten in biologischen Flüssigkeiten. Z. klin. Chem. klin. Biochem. *10:* 182–193 (1972).

9 NEUMEIER, D.; PRELLWITZ, W.; SANDEL, P. und KNEDEL, M.: Untersuchungen zur Aktivitätskinetik des Isoenzymes CK-MB im Serum nach Myokardinfarkt. Klin. Wschr. *56:* 449–456 (1978).

10 SHELL, W.E.; LAVELLE, J.F.; COVELL, J.W., and SOBEL, B.S.: Early estimation of myocardial damage in conscious dogs and patients with evolving acute myocardial infarction. J. clin. Invest. *52:* 2579–2590 (1973).

D. NEUMEIER, MD, Institut für Klinische Chemie, Klinikum Grosshadern, Universität München, Postfach 701 260, *D-8000 München 70* (FRG)

Enzymes in Health and Disease. Inaug. Scient. Meet. Int. Soc. Clin. Enzymol., London 1977, pp. 173–178 (Karger, Basel 1978)

Rapid and Quantitative Determination of Creatine Kinase MB Activity in Serum Using Immunological Inhibition of Creatine Kinase M Subunit Activity

U. WÜRZBURG, N. HENNRICH and H. LANG
Biochemical Research, E. Merck, Darmstadt

Introduction

Creatine kinase (CK; ATP: creatine phosphotransferase, EC 2.7.3.2) in the human organism occurs in the form of dimers of the subunits M and B, respectively. Therefore, 3 isoenzymes can be distinguished: CK-MM ('skeletal muscle type'), the hybrid CK-MB ('myocardial type'), and CK-BB ('brain type'). Recently, the measurement of CK-MB activity in serum has been shown to be a selective parameter of myocardial damage [1]. In the following, we describe a rapid and sensitive method for quantitative determination of CK isoenzyme MB activity. This method utilizes inhibiting antibodies (Inh-CK-M antibodies) which selectively block the activity of CK-M subunits. Since CK-BB activities only in very rare cases occur in the serum (estimated frequency about 1 : 10,000) [2, 3], for practical purposes all serum-CK-B subunit activity can be considered to be part of the CK-MB hybrid molecule. This CK-B activity is measured directly after immunological inhibition of all CK-M subunits present in the sample. The principle of this test is shown in figure 1. CK-MB activity can be calculated from the CK-B subunit activity in a simple way.

Experimental Procedures and Results

Preparation of antigens, antibodies, and test procedures were performed as described elsewhere [2, 4].

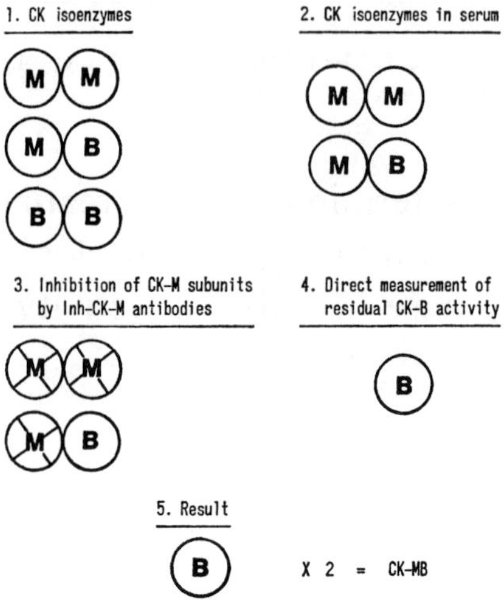

Fig. 1. Principle of the CK-MB determination by immunoinhibition. Schematic representation.

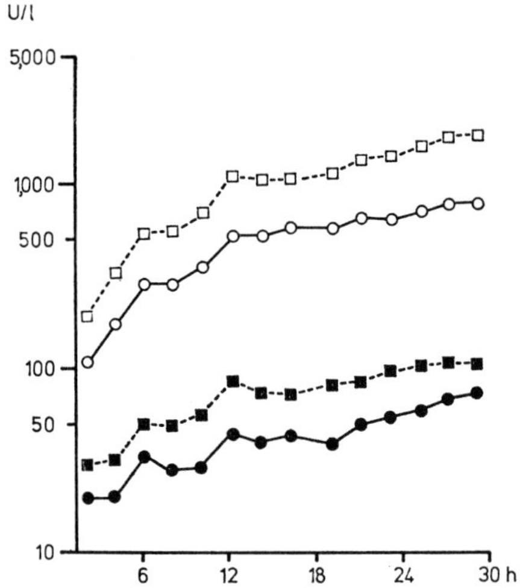

Fig. 2. Total serum CK activities and CK-MB activities in a patient with myocardial infarction. Simultaneous measurements with the glutathione (GSH) method [8] and the NAC method [9]. ○ = Total CK-GSH; □ = total CK-NAC; ● = CK-MB-GSH; ■ = CK-MB-NAC.

Table I. Inhibition of CK isoenzyme activities by inhibiting CK-M antibodies (goat). Inactivated human sera with addition of purified human CK isoenzymes. Mean ±1 SD from 5-fold determinations

Isoenzyme	Total activity after addition of human isoenzymes, U/l	Residual activity after incubation with Inh-CK-M antibodies, U/l
CK-MM	98±1.9	0.3±2.1
	1043±22	0.5±2.5
CK-MB	18±0.9	8.7±0.9
	37±0.7	18±0.8
	103±2.0	53±1.7
	410±7.8	206±6.2
CK-BB	197±3.8	199±4.1

Model Experiments

To demonstrate the degree and the specificity of inhibition of CK isoenzymes by Inh-CK-M antibodies from goats, experiments are performed on inactivated human sera, to which measured amounts of purified human CK isoenzymes are added. Table I contains the results of these measurements. Within the limits of error, the following percentage of activity inhibition is obtained: CK-MM=100%, CK-MB=50%, CK-BB=0%. These figures are valid over a wide range of activity, no differences were found in sera with CK activities between 10 and 1,000 U/l.

Comparable data according to the specificity and time course of inhibition were obtained in experiments with the new CK-NAC method at 37 °C and also at 25 °C [5, 6] following the recommendations of the SCE [7][1] and the German Society for Clinical Chemistry [9], respectively. Follow-up studies in sera of a patient with myocardial infarction (fig. 2) show good correlation of the different methods for CK activity measurements named above.

The influence of the animal species used to raise Inh-CK-M antibodies was shown previously [2]. Only goat antibodies yield 50% inhibition of CK-MB activity, rabbit and sheep antibodies inhibit by more than 50%,

[1] Scandinavian Committee on Enzymes of the Scandinavian Society for Clinical Chemistry and Clinical Physiology.

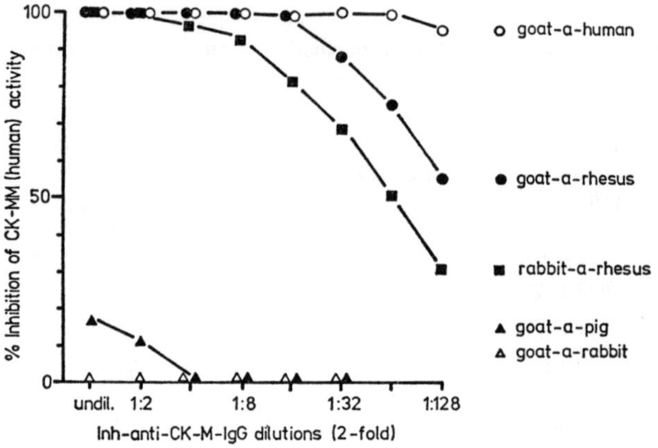

Fig. 3. Cross-inhibition of (human) CK-MM by inhibiting antibodies (Inh-anti-CK-M-IgG) raised against antigens from different species.

which means that besides CK-M subunit activity, also the B subunit activity in the CK-MB molecule is partially inhibited.

The influence of the antigen used for immunization on specificity of the resulting antibodies is shown in figure 3. The inhibition of human CK-MM activity is plotted against the antibody concentration. Besides human CK-MM, CK-MM from rhesus skeletal muscle also gives rise to antibodies in the goat, and in much lower concentration in rabbits, which inhibit human CK-MM by 100%. Antibodies raised against CK-MM from rabbit and pig muscle give no adequate cross-inhibition with the human isoenzyme.

The mechanism of CK-M inhibition under the described test conditions is not due to immunoprecipitation. This was tested by incubating CK-MM with Fab fragments of Inh-CK-M antibodies. These antibody fragments are monovalent and therefore cannot form immunoprecipitates. It was found that they exert the same inhibition characteristics as obtained with complete Inh-CK-M antibodies [2].

Conclusion

The Inh-CK-M antibodies from goats are incorporated into single vial reagents for CK activity measurements. With the aid of these reagents, quantitative determination of CK-MB activity can be performed by routine enzyme activity measurement within 15 min.

The results of CK-MB activity determinations in the sera of patients with different diseases using the immunoinhibition method are being published elsewhere [3, 4, 10–12].

Summary

A rapid and sensitive method for the quantitative determination of creatine kinase (CK) and isoenzyme MB (CK-MB) activity is presented. The method uses direct activity measurement of CK-B subunit activity after blocking of CK-M subunit activity by inhibiting antibodies. The test takes no longer than 15 min.

Inhibiting antibodies raised against CK-MM antigens of several species in different animals were characterized.

Data from model experiments with purified human CK isoenzymes demonstrate the specificity of activity inhibiting goat antibodies as independent from total CK activity and from different CK reagent systems.

References

1 Roberts, R. and Sobel, B. E.: CPK isoenzymes in evaluation of myocardial ischemic injury. Hosp. Pract. *11:* 55–62 (1976).
2 Würzburg, U.; Hennrich, N.; Orth, H. D.; Lang, H.; Prellwitz, W.; Neumeier, D.; Knedel, M., and Rick, W.: Quantitative determination of creatine kinase isoenzyme catalytic concentrations in serum using immunological methods. J. clin. Chem. clin. Biochem. *15:* 131–137 (1977).
3 Ljungdahl, L.; Hofvendahl, S.; Gerhard, W., and Börjesson, J.: Creatine kinase B-subunit activity in human serum. II. Evaluation of S-CK B-subunit activity in the diagnosis of acute myocardial infarction. Clin. chim. Acta *78:* 43–53 (1977).
4 Würzburg, U.; Hennrich, N.; Lang, H.; Prellwitz, W.; Neumeier, D. und Knedel, M.: Bestimmung der Aktivität von Creatinkinase MB im Serum unter Verwendung inhibierender Antikörper. Klin. Wschr. *54:* 357–360 (1976).
5 Gerhardt, W.; Ljungdahl, L.; Börjesson, J.; Hofvendahl, S., and Hedenäs, B.: Creatine kinase B-subunit activity in human serum. I. Development of an immunoinhibition method for routine determination of S-creatine kinase B-subunit activity. Clin. chim. Acta *78:* 29–41 (1977).
6 Würzburg, U. und Prellwitz, W.: Neue Ergebnisse bei Bestimmung der Aktivitäten von Creatinkinase-Isoenzymen mit Methoden auf immunologischer Basis. J. clin. Chem. clin. Biochem. *15:* 198 (1977).
7 The Scandinavian Committee on Enzymes: Scand. J. clin. Lab. Invest. *36:* 712–723 (1976).
8 Recommendations of the German Society for Clinical Chemistry: Standardisation of Methods for the Estimation of Enzyme Activities in Biological Fluids: J. clin. Chem. clin. Biochem. *8:* 658–660 (1970); J. clin. Chem. clin. Biochem. *10:* 182–192 (1972).

9 Recommendations of the German Society for Clinical Chemistry: J. clin. Chem. clin. Biochem. *12:* 249–254 (1977).

10 PRELLWITZ, W.; NEUMEIER, D.; KNEDEL, M.; LANG, H.; WÜRZBURG, U.; SCHÖNBORN, H. und SCHUSTER, H.P.: Isoenzyme der Kreatinkinase bei extrakardialen Erkrankungen und nach diagnostischen und therapeutischen Eingriffen: Dt. med. Wschr. *101:* 983–988 (1976).

11 PRELLWITZ, W. und NEUMEIER, D.: Such- und Schnellteste bei Herzerkrankungen und Kreislaufschock. Internist *17:* 36–444 (1976).

12 NEUMEIER, D.; PRELLWITZ, W.; WÜRZBURG, U.; BRUNDOBLER, M.; OLBERMANN, M.; JUST, H.-J.; KNEDEL, M., and LANG, H.: Determination of creatine kinase isoenzyme MB activity in serum using immunological inhibition of creatine kinase M subunit activity. Activity kinetics and diagnostic significance in myocardial infarction. Clin. chim. Acta *73:* 445–451 (1976).

H. LANG, PhD, Biochemical Research E. Merck, P.O. Box 4119, *D-6100 Darmstadt 1* (FRG)

Enzymes in Health and Disease. Inaug. Scient. Meet. Int. Soc. Clin. Enzymol., London 1977, pp. 179–185 (Karger, Basel 1978)

Lecithin-Cholesterol Acyltransferase (LCAT) Activity in Chronic Uremic Patients

G. F. Guarnieri, M. Moracchiello, L. Campanacci, F. Ursini, L. Ferri, M. Valente and C. Gregolin

Istituto di Patologia Medica dell'Università di Trieste, Trieste, and
Istituto di Chimica Biologica dell'Università di Padova, Padova

Introduction

Many atherogenic risk factors are present in chronic uremia, but lipid abnormalities are considered the most important ones [11, 15]. High incidence of hypertriglyceridemia has been reported in chronic uremic patients [1], with lipoprotein pattern abnormalities mainly resembling type IIa, IIb, III and IV hyperlipoproteinemia [4, 12, 17, 24]. A decrease of plasma high density lipoproteins (HDL) [16, 17] and low density lipoproteins 2 (LDL_2) [17], an increase of plasma very low density lipoproteins (VLDL) [17, 20] and LDL_1 [17], an impaired rate of VLDL catabolism and the presence of abnormal LDL [2, 17, 20, 24] have been described in chronic renal failure. Decreased triglyceride turnover [5] and plasma lipoprotein lipase (LPL) activity [1] and lower plasma phospholipids [14], cholesteryl esters [14, 17, 19], lysolecithin [22], polyunsaturated fatty acids [14] and erythrocyte cholesteryl esters [21] have also been reported in chronic uremic patients. An increased ratio of saturated/polyunsaturated fatty acids is also present in plasma cholesteryl esters and phospholipids of these patients [6, 14].

A most important contribution to some of these abnormalities might be an impaired activity of the plasma enzyme lecithin-cholesterol acyltransferase (LCAT) (EC 2.3.1.43). The physiological role of LCAT remains to be clarified. LCAT activity is considered to prevent unesterified cholesterol from accumulating in the plasma and to regulate lipoprotein catabolism [9, 10]. The enzyme, synthetized by the liver and secreted in the blood stream, acts almost exclusively on HDL. The cholesteryl esters formed can be transferred to VLDL and LDL in exchange for free cholesterol and triglycerides from VLDL to HDL [9, 10].

Table I. 'Intrinsic' and 'extrinsic' serum LCAT activity and serum lipids in chronic uremic patients

Patient	Sex	Total cholesterol mg/100 ml	Free/total cholesterol × 100	'Intrinsic' LCAT activity μg est. chol./ml serum/h	'Extrinsic' LCAT activity μg est. chol./ml serum/h	Cholesteryl ester clearance ml/h	Triglycerides mg/100 ml
1 I.C.	♀	82	31	26.6	19.5	113.2	–
2 G.Z.	♂	157	32	35.5	25.7	79.7	243
3 F.G.	♂	181	26	53.3	38.5	95.5	236
4 I.B.	♀	239	28	47.0	32.9	53.6	211
5 R.N.	♂	161	28	35.7	33.8	73.8	241
6 S.V.	♂	140	33	35.6	21.9	90.9	259
7 V.C.	♂	164	31	24.6	21.7	52.0	–
8 N.C.	♂	198	27	48.7	59.1	81.3	188
9 A.M.	♂	131	24	38.1	41.2	91.9	250
10 G.G.	♂	196	29	35.6	33.9	61.5	402
11 E.S.	♂	216	22	30.3	34.0	42.9	251
Mean ± SD		169[1] ± 43	28.27[1] ± 3.41	37.36[1] ± 9.04	32.93[1] ± 11.26	76.03 ± 21.61	253[1] ± 60
Normal subjects, mean ± SD		246 ± 12	22.06 ± 1.80	54.57 ± 5.20	45.13 ± 2.98	68.06 ± 2.86	120 ± 17

[1] $p < 0.01$, as compared to control values (Student's *t*-test for unpaired data).

In this work, LCAT activity has been determined in serum of chronic uremic patients on conservative treatment. Patients on maintenance hemodialysis were excluded from this study, owing to the repeated exposure to heparin infusions, which can stimulate VLDL catabolism and decrease LCAT activity [25]. The activity was determined by using the serum of each patient both as a source of enzyme and as a substrate ('intrinsic' activity) [8] and was compared with the activity determined on a standard substrate ('extrinsic' activity) [23], so as to ascertain the presence of inhibitory factors in the patients' serum. While the intrinsic LCAT activity is a measure of the actual activity in circulating plasma, the extrinsic one is an index of the enzyme concentration.

Both intrinsic and extrinsic activities have been found significantly decreased in chronic uremia.

Methods

Eleven patients with chronic renal failure, aged 27–65, taken at random, were included in this study. Patients affected by diabetes, hepatic insufficiency, congestive heart failure, or nephrotic syndrome were excluded. Diet and nutritional status of the patients were not uniform. They were generally maintained on a normocaloric (30–35 kcal/kg/day) and low protein (22–40 g/day) diet. Serum creatinine was determined by a Technicon autoanalyser; its range was 3.1–15 mg%. Ten controls (aged 32–50) were selected. All controls had normal renal function and no alteration in carbohydrate or lipid metabolism.

Fasting blood samples were obtained from all subjects; after clotting, the sera were frozen at –20 °C or immediately used for determinations. Serum triglycerides were determined by Eggstein's method [7]. Total cholesterol and free cholesterol (after precipitation as digitonin complex) were determined by the Liebermann-Buchard reaction.

LCAT activity was determined by a modification of Glomset's method [8, 23].

The clearance of cholesteryl esters was also measured, i.e. the volume of serum from which cholesteryl esters are removed in a given time, to maintain a steady cholesterol/cholesteryl esters ratio in the presence of continuous cholesterol input and esterification in blood [8].

Student's *t*-test for unpaired data and linear regression analysis were used for statistical calculations.

Results

Table I summarizes the results of the determinations of serum total and free cholesterol, triglycerides, intrinsic and extrinsic LCAT activity and cholesteryl ester clearance in uremic patients and in controls.

In patients' serum, the triglyceride concentration was significantly higher and the total cholesterol concentration significantly lower than in controls' serum, with an increased percentage of the unesterified form. The average LCAT activity (both intrinsic and extrinsic) in uremic patients was significantly lower than in the controls. The intrinsic LCAT activity was higher than the extrinsic both in patients and in controls, except in 3 cases. When analyzed by linear regression, there was a significant correlation ($r = 0.69$, $p < 0.05$) between intrinsic and extrinsic LCAT activities

Discussion

The low concentration of plasma cholesteryl esters [14, 17, 19] and lysolecithin [22], the increased concentration of VLDL [17, 20] and the presence of abnormal LDL in uremic sera [2, 17, 20, 24] are consistent with a low LCAT activity. Actually, both intrinsic and extrinsic enzyme activities have been found significantly decreased. The two activities varied in parallel, supporting the concept that no factor in the uremic plasma inhibits the enzyme activity, nor a relative unavailability of some substrates in chronic uremia (e.g. polyenoic fatty acids, phospholipids, HDL) is relevant.

A low extrinsic activity suggests that in chronic uremic patients, the enzyme synthesis in the liver is probably reduced, owing to a lower level of circulating cholesterol. Moreover, LCAT activity has been found to parallel the levels of plasma proteins in liver disease [3], and the hepatic synthesis of many plasma proteins is actually reduced in chronic uremic patients [13].

The intrinsic LCAT activity was generally higher than the extrinsic, both in patients and in controls, probably owing to partial alterations of the lipoproteins during the heating of the serum substrate used for the extrinsic activity test.

This reduced LCAT activity ensures a cholesteryl ester clearance in the range of normal subjects. This is surprising, if related to the enhanced triglyceride concentration, because in other forms of hypertriglyceridemia an enhancement of LCAT activity, proportional to triglyceride content, as well as an increase of the cholesteryl ester clearance, were generally found [8, 23, 25]. It can be argued that in chronic uremia the VLDL (i.e. the triglycerides) do not properly stimulate the LCAT activity and the cholesteryl ester clearance and, as a consequence, the VLDL fail to promote their own catabolism. The fact that the percentage of unesterified cholesterol is augmented in chronic uremia is in line with this argument.

The raised levels of VLDL [17] and free cholesterol [18], associated with the decrease of LCAT activity, may be a contributing factor to the accelerated development of atherosclerotic lesions in chronic uremia.

Summary

Serum lecithin-cholesterol acyltransferase (LCAT) activity and cholesteryl ester clearance were determined in eleven patients with chronic renal failure and in ten controls. LCAT activity has been found significantly lower in chronic uremic patients than in controls. The cholesteryl ester clearance did not apparently respond to the stimulatory effect of hypertriglyceridemia, as observed in other cases of dislipoproteinemias. The presence of 'uremic toxins', inhibiting the enzyme activity, is unlikely. LCAT synthesis in the liver is probably reduced in chronic uremia.

Acknowledgements

A preliminary report of this work was presented at the Congress of the Italian Society of Nephrology, Rapallo, 1976 (Kidney int. *11:* 141, 1977).

The authors wish to thank Mrs. Guiliana Giungarelli and Miss Fedora Tamburini for secretarial assistance.

This work was supported by Consiglio Nazionale delle Ricerche Roma, Grants No. 72–00342/04 115–3622 and 76–01348.04.115.1111.

References

1 Bagdade, J.D.: Lipemia, a sequela of chronic renal failure and hemodialysis. Am. J. clin. Nutr. *21:* 426–429 (1968).

2 Bagdade, J.; Casaretto, A., and Albers, J.: Effects of chronic uremia, hemodialysis, and renal transplantation on plasma lipids and lipoproteins in man. J. Lab. clin. Med. *87:* 37–48 (1976).

3 Blomhoff, J.P.; Skrede, S., and Ritland, S.: Lecithin: cholesterol acyltransferase and plasma proteins in liver disease. Clin. chim. Acta *53:* 197–207 (1974).

4 Brøns, M.; Christensen, N.C., and Hørder, M.: Hyperlipoproteinemia in patients with chronic renal failure. Acta med. scand. *192:* 119–123 (1972).

5 Cattran, D.C.; Steiner, G.; Fenton, S.S.A., and Wilson, D.R.: Hypertriglyceridemia in uremia and the use of triglyceride turnover to define pathogenesis. Trans. Am. Soc. artif. internal Organs *20:* 148–152 (1974).

6 Ceccanti, M.; Chiarioni, T.; Romeo, M.; Santini, P.; Sasso, G.F. e Tagliamonte, L.: Modificazioni dei lipidi plasmatici nell'insufficienza renale cronica; in Atti 77° Congr. Soc. It. Med. Int., pp. 179–185 (Pozzi, Roma 1977).

7 Eggstein, M.: Eine neue Bestimmung der Neutralfette im Blutserum und Gewebe. II. Mitteilung. Klin. Wschr. *44:* 267–273 (1966).

8 Ferri, L.; Ursini, F.; Valente, M., and Gregolin, C.: Lecithin-cholesterol acyltransferase (LCAT) activity as a determining factor in lipoprotein catabolism: a study on hyperlipoproteinemic subjects; in Burlina and Galzigna Proc. 7th Int. Symp. on Clin. Enzymol., pp. 125–137 (Piccin, Padua 1977).

9 Glomset, J.A. and Norum, K.A.: The metabolic role of lecithin:cholesterol acyltransferase: perspectives from pathology. Adv. Lipid Res. *11:* 1–63 (1973).

10 Glomset, J.A.; Nichols, A.V.; Norum, K.A.; King, W., and Forte, T.: Plasma lipoproteins in familial lecithin: cholesterol acyltransferase deficiency. Further studies of very low and low density lipoprotein abnormalities. J. clin. Invest. *52:* 1078–1092 (1973).

11 Guarnieri, G.F.: Ipertrigliceridemia ed accelerata arteriosclerosi nell'insufficienza renale cronica. G. ital. Clin. Chim. *1:* 47–61 (1976).

12 Hübner, A.; Sieberth, H.G.; Diemer, A.; Finke, K., and Prange, E.: Effects of regular haemodialysis with glucose and glucose free dyalisate on hyperlipemia; in Cameron Proc. Eur. Dial. Transpl. Ass., vol. VIII, pp. 174–181 (Pitman, London 1971).

13 Kopple, J.D.: Dietary requirements; in Massry and Sellers Clinical aspects of uremia and dialysis, pp. 453–489 (Thomas, Springfield 1976).

14 Lambert, E.L., jr. and Oates, J.A.: Plasma lipid patterns in chronic renal failure. Trans. Am. Soc. artif. internal Organs *21:* 473–477 (1975).

15 Lazarus, J.M.; Lowrie, E.G.; Hampers, C.L., and Merrill, J.P.: Cardiovascular disease in uremic patients on hemodialysis. Kidney int. *7:* suppl. 2, 167–175 (1975).

16 Lewis, L.A.; Zuehlke, V.; Nakamoto, S.; Kolff, W.J., and Page, I.H.: Renal regulation of serum alpha-lipoproteins. Decrease of alpha-lipoproteins in the absence of renal function. New Engl. J. Med. *275:* 1097–1100 (1966).

17 Norbeck, H.-E.; Orö, L., and Carlson, L.A.: Serum lipid and lipoprotein concentrations in chronic uremia. Acta med. scand. *200:* 487–492 (1976).

18 Portman, O.W.; Alexandre, M., and Osuga, T.: Heterogeneity of lipid composition of microsome subfractions from aorta and liver. Biochim. biophys. Acta *187:* 435–438 (1969).

19 Reimold, W.V.; Kattermann, R. und Quellhorst, E.: Serumlipide bei chronischer Niereninsuffizienz und deren Beeinflussung durch Glukose und Sorbit während der Peritonealdialyse. Arch. klin. Med. *216:* 384–401 (1969).

20 Samar, R.E.; Moncrief, J.W.; Decherd, J.F., and Popovich, R.P.: Lipoprotein binding and hypertriglyceridemia in chronic uremia. Trans. Am. Soc. artif. internal Organs *21:* 455–462 (1975).

21 Sasso, G.F.; Romeo, M.; Nardi, E.; Grasso, P.; Chiarioni, T. e Ceccanti, M.: Modificazioni dei lipidi eritrocitari nell'insufficienza renale cronica; in Atti 77° Congr. Soc. It. Med. Int., pp. 186–189 (Pozzi, Roma 1977).

22 Schardt, W. and Wardle, E.N.: Plasma lysolecithin in uraemia. Clin. chim. Acta *63:* 223–225 (1975).

23 Ursini, F.; Ferri, L.; Esposito, C.; Valente, M.; Gregolin, C.; Cortesi, S., and Rigoni, G.: Role of the LCAT reaction in diseases with serum lipid perturbation; in Burlina Proc. of the 6th Symp. on Clin. Enzymol., pp. 439–459 (Kurtis, Milano 1974).

24 Wada, M.; Minamisono, T.; Fujii, H.; Morita, T.; Akamatsu, A.; Mise, J.; Nakamoto, S., and Naito, H.K.: Studies on the effects of hemodialysis on plasma lipoproteins. Trans. Am. Soc. artif. internal Organs *21:* 464–471 (1971).
25 Wallentin, L. and Vikrot, O.: Lecithin:cholesterol acyl transfer in plasma of normal persons in relation to lipid and lipoprotein concentration. Scand. J. clin. Lab. Invest. *35:* 669–676 (1975).

G.F. Guarnieri, MD, Istituto di Patologia Medica dell'Università di Trieste, c/o Ospedale Maggiore, Piazza Ospedale No.1, *I-34129 Trieste* (Italy)

Enzymes in Health and Disease. Inaug. Scient. Meet. Int. Soc. Clin. Enzymol., London 1977, pp. 186–192 (Karger, Basel 1978)

Urinary Enzymes (α-Glucosidase and Lysozyme), Protein Pattern and β_2-Microglobulin as Indexes of Tubular Damage

I. After Cardiac Surgery

G. F. Guarnieri, L. Faccini, E. Beltram, M. Chersicla, G. Battilana, B. Branchini, P. Ferrara, W. Micheli, S. Lin and L. Campanacci[1]

Istituto di Patologia Medica dell'Università di Trieste; Divisione di Cardiochirurgia, and Servizio di Medicina Nucleare, Ospedale Maggiore, Trieste

Introduction

Urinary enzymes are proving to be useful diagnostic tools [5–7, 9, 13] especially when rightly classified – according to other urinary proteins – as indexes of glomerular, tubular and parenchymatous enzymuria [3–7]. α-Glucosidase (EC 3.2.1.20) is a lysosomal enzyme present in the renal tubular epithelium and absent in plasma, which increases in the urine in consequence of tubular damage ('parenchymatous enzymuria') [3–7]. Lysozyme (muramidase) (EC 3.2.1.17) is a microprotein present in plasma, which normally filters through the glomeruli and is reabsorbed by proximal tubular cells. It also increases in the urine in consequence of tubular damage and is generally considered an index of tubular enzymuria [5–7, 11–13]. The combined study of parenchymatous and tubular enzymuria seems to be the most useful one in various renal diseases [5–7]. In the present study, the urinary excretion of α-glucosidase and lysozyme has been investigated after cardiac operations with cardiopulmonary bypass, which often result in kidney damage. The enzymuria has been compared with the urinary excretion of β_2-microglobulin, which is considered an index of tubular proteinuria [14] and with the electrophoretic pattern of urinary proteins.

[1] With the technical assistance of P. Crismani and B. Pierini.

Materials and Methods

The number of patients examined is reported in table I. Details of the surgical procedures – performed according to standard techniques – are reported elsewhere [10]. α-Glucosidase has been determined by the method of FRANZINI *et al.* [6], lysozyme by the method of PROCKOP and DAVIDSON [10, 12]. The extinction change of the lysozyme reaction was read in an Eppendorf photometer with automatic recorder. Creatinine has been determined by the Biochemica Test Combination from Boehringer and β_2-microglobulin by the Phadebas β_2-micro test from Pharmacia. Urinary protein electrophoresis was performed on Cellogel RS strips, following the standard Chemetron method. Results have been expressed as 'fractional' α-glucosidase excretion and 'fractional' lysozyme and β_2-microglobulin clearances, by relating enzymuria to creatinine clearance [5–7, 11]. By this means, variations due to plasma concentration, urinary flow and mass of functioning nephrons are eliminated. Wilcoxon's test for paired data and Spearman's rank coefficient were used for statistical calculations.

Results and Discussion

After cardiac surgery with cardiopulmonary bypass (CPB), we observed a statistically significant two-phase increase of mean fractional α-glucosidase excretion (table I).

Mean fractional lysozyme clearance significantly increased only immediately after surgery.

β_2-Microglobulin was also significantly higher after cardiac surgery.

After cardiac operations, urinary α-glucosidase increased in 23 out of 26 patients (mean increase 532%, range 35–3,091%), lysozyme in 21 out of 27 patients (mean increase 6,073%, range 150–44,025%) and β_2-microglobulin in 8 out of 9 patients (mean increase 6,248%, range 103–25,600%). During the 5 days after surgery, an increase of serum creatinine concentration of at least 0.5 mg/dl was detected in 10 out of 16 patients. On the 5th day after surgery, urinary α-glucosidase was still increased in 12 out of 16 patients (mean increase 376%, range 28–934%) and urinary lysozyme in 9 out of 16 patients (mean increase 3,217%, range 62–24,400%). The electrophoretic pattern of proteinuria, studied in 17 patients, was normal in everyone before cardiac surgery. After operation, a tubular proteinuria pattern generally appeared, which often changed to a mixed pattern on the following days.

Cardiac operations without CPB have been performed in 4 other patients. Immediately after surgery, urinary enzymes increased only in 1 of these patients. No tubular proteinuria appeared, whereas a glomerular pattern was detectable in 2 patients on the following days.

Table I. Effects of CPB on urinary enzymes and proteins

	Before	After	Day 1	Day 3–4	Day 5–6
$\frac{\text{Urinary } \alpha\text{-glucosidase, mU/min}}{\text{Creatinine clearance, ml/min}} \times 100$	1.40±0.47 (16)	3.95±0.64 ≦0.01[1]	2.13±0.29 n.s.	3.30±0.49 ≦0.01	2.12±0.24 n.s.
	1.38±0.30 (26)	4.72±0.85 ≦0.01			
$\frac{\text{Lysozyme clearance, ml/min}}{\text{Creatinine clearance, ml/min}} \times 100$	0.85±0.69 (16)	2.99±1.34 n.s.	1.14±0.50 n.s.	0.60±0.35 n.s.	0.92±0.48 n.s.
	0.55±0.41 (27)	3.32±1.02 ≦0.01			
$\frac{\beta_2\text{-Microglobulin clearance, ml/min}}{\text{Creatinine clearance, ml/min}} \times 100$	0.36±0.49 (8)	4.46±3.71 ≦0.02	3.83±5.17 ≦0.02	2.04±4.25 ≦0.05	0.88±0.53 ≦0.01
	0.33±0.47 (9)	4.32±3.50 ≦0.01			

Pattern of the proteinuria[2]

Patient					
G.P.	N	N	N		
R.B.	N	N	N	N	G
A.G.	N	N	T		
R.L.	N	N	T	T	N
C.D.	N	T	N		
D.T.	N	T	N	N	N
C.D.	N	T	N	T	N
S.B.	N	T		N	G
F.R.	N	T	T	N	
G.M.	N	T	M	M	M
S.C.	N	T	M	M	M
E.C.	N	T	M	N	M
E.S.	N	T	M	M	G
D.D.	N	T	M	M	M
S.O.	N		M		M
G.P.	N	T			

Numbers in parentheses represent number of patients. Two sets of numerical data are presented as mean±SEM, the upper representing patients followed sequentially until day 5–6, and the lower those patients in whom one set of data after CPB was available. The pattern of proteinuria is documented in 16 patients.

[1] Wilcoxon's test for paired data vs basal values. Numbers refer to level of significance.

[2] N=Normal; T=tubular; G=glomerular; M=mixed (glomerular and tubular).

By determining Spearman's rank coefficient, a significant correlation was found between duration of CPB and both fractional α-glucosidase excretion ($r' = 0.44$, level of significance 5%) and fractional lysozyme clearance ($r' = 0.56$, level of significance 1%).

Mean lysozymuria was higher in patients treated with gentamycin after surgery (1.21 ± 2.25) than in patients not treated with this antibiotic (0.28 ± 0.48).

Renal failure seems to occur more frequently after cardiovascular surgery than after other surgical procedures and its occurrence carries a severe prognosis [1, 2, 8]. Owing to this high mortality rate, it would be important to predict and possibly to prevent those risk factors which lead to the development of acute renal failure. The pathogenesis is probably multifactorial and is related to preoperative, intraoperative and post-operative factors [1, 2, 8]. Renal failure developing in the immediate post-operative period is considered to be specifically related to open heart surgery (e.g. the non-pulsatile flow pattern of bypass, leading to renal hemodynamic abnormalities), whereas renal failure appearing later in the postoperative period probably depends on various superimposed complications [2]. The high frequency of renal abnormalities after cardiovascular surgery with CPB is evident from our data. Urinary enzyme and protein determinations may be regarded as sensitive predictors of both preoperative risk factors and postoperative complications. Still more important could be their role in recognizing the pathogenesis of renal failure after cardiac operations. The increase of urinary enzymes and β_2-microglobulin and the frequent appearance of a tubular proteinuria pattern immediately after surgery with CPB, and the correlation of enzymuria with duration of CPB, suggest an influence of CPB itself in altering the renal function. The late increase of enzymuria and the late appearance of a glomerular proteinuria pattern probably reflect a kidney damage not specifically related to cardiac surgery. The appearance of a glomerular proteinuria with circulatory disturbances is a well-known phenomenon. As in our series, an increase of urinary lysozyme and protein fractions has been reported 3–4 days after various surgical procedures [15]. Our results suggest that the administration of nephrotoxic drugs, such as gentamycin, may be relevant too. In previous studies, we observed an increase of urinary α-glucosidase after treatment with gentamycin in normal subjects too [5, 6].

These results suggest that the determination of critically selected urinary enzymes, together with the determination of urinary protein pattern, may be valuable both for diagnostic purposes and for pathogenetic investigations.

Summary

A significant increase of urinary α-glucosidase, lysozyme and β_2-microglobulin and the appearance of a tubular and a mixed proteinuria pattern were observed after cardiac surgery with cardiopulmonary bypass. These results indicate tubular damage, resulting both from cardiopulmonary bypass and postoperative complications.

Acknowledgements

A preliminary report of this work was presented at the XVIII Congress of the Italian Society of Nephrology, Ancona, 1977.

The authors wish to thank Miss Fedora Tamburini for secretarial assistance.

References

1 Abel, R.M.; Buckley, M.J.; Austen, W.G.; Barnett, G.O.; Beck, C.H., jr., and Fischer, J.E.: Etiology, incidence, and prognosis of renal failure following cardiac operations. Results of a prospective analysis of 500 consecutive patients. J. thorac. cardiovasc. Surg. *71:* 323–333 (1976).

2 Bhat, J.G.; Gluck, M.C.; Löwenstein, J., and Baldwin, D.S.: Renal failure after open heart surgery. Ann. intern. Med. *84:* 677–682 (1976).

3 Bonini, P.A. and Ceriotti, G.: Human urinary α-glucosidase as an index of kidney tubular damage. Clin. chim. Acta *27:* 415–419 (1970).

4 Bonini, P.A.; Guarnieri, G.F.; Nadai-Frank, A. de e Ceriotti, G.: Significato fisiopatologico e applicazioni cliniche degli enzimi urinari. Quad. Sclavo Diagn. *9:* 419–441 (1973).

5 Campanacci, L.; Guarnieri, G.F.; Faccini, L.; Bellini, G. e Carretta, R.: Utilità di alcune indagini enzimologiche in nefrologia clinica. Min. Nefrol. *22:* 155–169 (1975).

6 Ceriotti, G. and Guarnieri, G.F.: Physiological background for a possible utilization of urinary enzymes in clinical diagnosis; in Burlina Proc. 6th Int. Symp. on Clin. Enz., pp. 255–270 (Kurtis, Milano 1974).

7 Ceriotti, G. e Guarnieri, G.F.: L'interesse diagnostico della determinazione degli enzimi urinari. Quad. Sclavo Diagn. *11:* 103–121 (1975).

8 Chesney, R.W.; Kaplan, B.S.; Freedom, R.M.; Haller, J.A., and Drummond, K.N.: Acute renal failure: an important complication of cardiac surgery in infants. J. Pediat. *87:* 381–388 (1975).

9 Dubach, U.C.: Enzymes in urine and kidney. Proc. Int. Conf. on Enzymes in Urine, Rheinfelden 1967 (Huber, Berne 1968).

10 Guarnieri, G.F.; Faccini, L.; Beltram, E.; Chersicla, M.; Battilana, G.; Milani, S.; Lin, S. e Campanacci, L.: Indagini enzimologiche sull'integrità del tubulo renale in corso di polichemioterapia antiblastica, dopo somministrazione e.v. die mezzi di contrasto ed in operati in circolazione extracorporea. Atti XVIII Congr. Soc. It. Nefrol. Ancona 1977 (in press).

11 HARRISON, J.F.; LUNT, G.S.; SCOTT, P., and BLAINEY, J.D.: Urinary lysozyme, ribonuclease, and low-molecular-weight protein in renal disease. Lancet *i:* 371–375 (1968).
12 PROCKOP, D.J. and DAVIDSON, W.D.: A study of urinary and serum lysozyme in patients with renal disease. New Engl. J. Med. *270:* 269–274 (1964).
13 RAAB, W.P.: Diagnostic value of urinary enzyme determinations. Clin. Chem. *18:* 5–25 (1972).
14 REVILLARD, J.P. et VINCENT, C.: La β_2-microglobuline. Nouv. Presse méd. *5:* 2707–2712 (1976).
15 SKURZEWSKI, J.; SZEWCZYK, Z.; MELCER, H.; KOPEC, W., and GALICKA, G.: Urine excretion of muramidase, protein and protein fractions in patients in early post-operative period. 12th Conf. Eur. Dial. and Transplantn Ass., abstr., p. 213 (1975).

G.F. GUARNIERI, MD, Istituto di Patologia Medica dell'Università di Trieste, c/o Ospedale Maggiore, Piazza Ospedale 1, *I-34129 Trieste* (Italy)

Enzymes in Health and Disease. Inaug. Scient. Meet. Int. Soc. Clin. Enzymol., London 1977, pp. 193–200 (Karger, Basel 1978)

Urinary Enzymes (α-Glucosidase and Lysozyme), Protein Pattern and β_2-Microglobulin as Indexes of Tubular Damage

II. After Administration of Cytostatic Agents and Contrast Media

G. F. GUARNIERI, L. FACCINI, M. CHERSICLA, E. BELTRAM, G. BATTILANA, S. MILANI, M. BAZZOCCHI, M. P. MUCCI, S. LIN and L. CAMPANACCI[1]

Istituto di Patologia Medica dell'Università di Trieste; Istituto di Radiologia dell'Università di Trieste, and Servizio di Medicina Nucleare, Ospedale Maggiore, Trieste

Introduction

Urinary enzymes may be regarded as a sensitive index of tubular damage [2, 3, 5–7, 10, 13, 15, 20, 21].

In the present study, the urinary excretion of α-glucosidase (EC 3.2.1.20) and lysozyme (muramidase) (EC 3.2.1.17) [13] has been investigated after some procedures, resulting sometimes in kidney damage, such as the administration of potentially nephrotoxic agents (radiological contrast media, cytostatic agents). The enzymuria has been compared with the urinary excretion of β_2-microglobulin, and with the electrophoretic pattern of urinary proteins [13].

Material and Methods

The number of patients examined is reported in tables I–III. Details of the diagnostic and therapeutic procedures – performed according to standard techniques – as well as the methods of analysis are reported elsewhere [8, 13, 14].

[1] With the technical assistance of P. CRISMANI and B. PIERINI

Table I. Effects of high dose urography with 'Conray 400' on urinary enzymes and proteins

	Before	After	Day 1–2	Day 3–4	Day 5–6
$\frac{\text{Urinary } \alpha\text{-glucosidase, mU/min}}{\text{Creatinine clearance, ml/min}} \times 100$	2.91 ± 0.76 (8)	3.10 ± 0.85 n.s.[1]	3.42 ± 0.81 n.s.	1.91 ± 0.29 n.s.	3.84 ± 1.05 n.s.
	2.38 ± 0.47 (18)	2.62 ± 0.64 n.s.			
$\frac{\text{Lysozyme clearance, ml/min}}{\text{Creatinine clearance, ml/min}} \times 100$	0.60 ± 0.50 (8)	1.10 ± 0.54 n.s.	1.02 ± 0.82 n.s.	1.30 ± 1.00 n.s.	1.27 ± 0.59 n.s.
	1.46 ± 0.97 (13)	2.19 ± 0.78 n.s.			
$\frac{\beta_2\text{-Microglobulin clearance, ml/min}}{\text{Creatinine clearance, ml/min}} \times 100$	1.38 ± 1.24 (4)	0.50 ± 0.35 n.s.	1.69 ± 1.66 n.s.	1.38 ± 1.31 n.s.	
Pattern of the proteinuria[2]					
Patient					
Z.B.	N	N	N	N	N
M.P.	N	N	N	N	N
C.M.	N	N	N	N	M
M.S.	N	G	G	G	G
G.K.	N	T	M	M	M
Y.P.	M	M	M	M	M
B.C.	M	M	M	M	M
O.F.	T	M	M	N	N

Numbers in parentheses represent number of patients. Two sets of numerical data are presented as mean ± SEM for fractional α-glucosidase excretion and lysozyme clearance, the upper representing patients followed sequentially until day 5–6, and the lower those patients in whom one set of data after urography was available. β_2-Microglobulin clearance is presented as mean ± SEM for 4 patients examined sequentially up to day 3–4. The pattern of proteinuria is documented in 8 patients.

[1] Wilcoxon's test for paired data vs basal values.

[2] N = Normal; T = tubular; G = glomerular; M = mixed (glomerular and tubular).

Results and Discussion

Effects of high dose urography (HDU) (table I). A two-phase increase of mean fractional α-glucosidase excretion and an increase of mean fractional lysozyme clearance were observed, but variations were statistically not significant. However, it is to be pointed out that an increase of α-glucosidase in 6 out of 18 patients (mean increase 95%, range 35–952%), of lysozyme in 11 out of 13 patients (mean increase 2,252%, range 69–10,700%) and β_2-microglobulin in 1 out of 4 patients (63%) has been detected.

On the 5th day after HDU, an increase of urinary α-glucosidase in 3 patients (33,145 and 356%) and of urinary lysozyme in 6 patients (mean increase 1,607%, range 54–5,000%) was still present. We observed the appearance of a tubular proteinuria in 1 patient, of a glomerular proteinuria in 2 others and of a mixed proteinuria in 1. Serum creatinine increased only in 1 patient (0.6 mg/100 ml).

A transient decrease in renal function – as evaluated by serum creatinine content and creatinine clearance – because of a nephrotoxic effect of contrast media has been reported in some patients after HDU [17, 18]. Bengtsson *et al.* [1], however, did not observe any significant change in inulin and PAH clearances after HDU. A transient, but definite increase of lysozymuria in dogs [9] and of alaninaminopeptidase in men [4] has been found after radiological contrast media administration. These results, as well as those reported in the present study, suggest that urinary enzymes are a more sensitive and specific index of kidney damage than serum creatinine determination and renal clearance techniques. Their routine use in high risk patients, such as those with multiple myeloma, diabetes mellitus and fluid restriction [18], could allow early diagnosis of tubular damage. Further studies are necessary to demonstrate a possible higher toxicity of contrast media in patients with severe impairment of renal function.

Effects of arteriography (table II). A two-phase increase of mean fractional α-glucosidase excretion and a transient increase of fractional lysozyme clearance were also observed after arteriography. In the larger group of patients studied only before and after arteriography the increase of both fractional α-glucosidase excretion and lysozyme clearance was statistically significant. Urinary α-glucosidase increased in 11 out of 17 patients (mean increase 270%, range 31–1,024%) and lysozyme in 16 out of 17 patients (mean increase 1,179%, range 100–5,900%). Mean β_2-microglobulin clearance also increased after arteriography in 4 out of 5 patients (975–4,330–11,100–18,500%). On the 5th day after arteriography, an increase of urinary

Table II. Effects of arteriography with 'Urografin' on urinary enzymes and proteins

	Before	After	Day 1	Day 3	Day 5–6
$\frac{\text{Urinary } \alpha\text{-glucosidase, mU/min}}{\text{Creatinine clearance, ml/min}} \times 100$	1.66 ± 0.32 (7)	2.43 ± 0.63 n.s.[1]	1.83 ± 0.38 n.s.	1.00 ± 0.14 n.s.	1.88 ± 0.51 n.s .
	1.58 ± 0.13 (17)	2.74 ± 0.37 ≦0.01			
$\frac{\text{Lysozyme clearance, ml/min}}{\text{Creatinine clearance, ml/min}} \times 100$	0.05 ± 0.02 (7)	0.42 ± 0.20 n.s.	0.05 ± 0.03 n.s.	0.09 ± 0.03 n.s.	0.06 ± 0.02 n.s.
	0.23 ± 0.13 (17)	3.44 ± 2.86 ≦0.01			
$\frac{\beta_2\text{-Microglobulin clearance, ml/min}}{\text{Creatinine clearance, ml/min}} \times 100$	0.11 ± 0.07 (5)	1.58 ± 0.78 n.s.	0.05 ± 0.03 n.s.	0.03 ± 0.01 n.s.	
Pattern of the proteinuria[2]					
Patient					
P.M.	N	N	N	N	N
A.G.	N	N		N	N
M.C.	N	N	G	N	
E.S.	N	G	N	N	
A.D.	N	G	G	G	N
A.M.	N	G		G	
L.M.	G[3]	G	G	G	N
M.S.	G[3]	G		G	G

Numbers in parentheses represent number of patients. Two sets of numerical data are presented as mean ± SEM for fractional α-glucosidase excretion and lysozyme clearance, the upper representing patients followed sequentially until day 5–6, and the lower those patients in whom one set of data after urography was available. β_2-Microglobulin clearance is presented as mean ± SEM for 5 patients examined sequentially up to day 3. The pattern of proteinuria is documented in 8 patients.

[1] Wilcoxon's test for paired data vs basal values. Numbers refer to level of significance.

[2] N = Normal; G = glomerular.

[3] Arteriography was performed few days after high dose urography. Urinary protein pattern was normal before high-dose urography.

α-glucosidase in 1 patient (91%) and of lysozyme in 2 patients (200 and 800%) was still detectable. Following arteriography, a glomerular proteinuria sometimes appeared. It is interesting that in patients L.M. and M.S., a glomerular urinary protein pattern had appeared as a consequence of HDU and was still present when arteriography was performed. Permanent renal impairment has been described in 2 patients who underwent renal arteriography before renal function had been restored after urography [18]. We observed no increase of serum creatinine content in our patients after arteriography.

Proximal tubular necrosis has been described after angiography [11, 16]. Tubular damage in 6 out of 17 patients following selective renal angiography has been reported by GOLDSTEIN *et al.* [12], who observed the appearance of ligandin – a glutathione transferase present in proximal tubular cells – in the urine of these patients. Our results suggest that the appearance of urinary abnormalities is frequent after arteriography. This points to tubular damage, which is generally transient and slight. Routine prolonged determinations of urinary enzymes seem worth being performed in high-risk patients [18, 19] owing to the much higher sensitivity of enzymuria in detecting tubular damage, in comparison with usual kidney function tests. We are now investigating if a dose-related nephrotoxic effect of contrast media is detectable and if a higher nephrotoxicity can be demonstrated after direct injection of contrast media in the renal artery (selective renal angiography).

Effects of the administration of cytostatic agents (table III). Mean urinary α-glucosidase excretion significantly increased after the administration of the two doses of methotrexate and 5-fluorouracil. Urinary α-glucosidase increased in 9 out of 22 patients (mean increase 79%, range 41–148%) after the first administration and in 6 out of 18 patients (mean increase 305%, range 53–1,167%) after the second administration. No statistically significant variation of lysozyme excretion was detected, although an increase of lysozymuria was observed in 4 out of 17 patients (107, 277, 289 and 1,231%) after the first administration and in 7 out of 13 patients (mean increase 591%, range 80–3,050%) after the second administration. At the end of the cycle an increase of urinary α-glucosidase in 4 patients (31, 45, 60, 158%) and of urinary lysozyme in 5 patients (56, 74, 173, 212 and 964%) was still detectable.

A nephrotoxic effect of antitumor drugs has been recently reported in animals [22] and a transient cytotoxic effect of these drugs on proximal tubular cells is also evident from our study. It cannot be excluded that

Table III. Effects of the administration of cytostatic agents[1] on urinary enzymes

	Day 0	Day 1	Day 3	Day 7	Day 8	Day 10
$\frac{\text{Urinary } \alpha\text{-glucosidase}}{\text{Urinary creatinine}}$, mU/mg	0.99±0.34 (12)	1.22±0.39 ≦0.01[2]	1.18±0.59 n.s.[2]	0.94±0.46	1.30±0.26 ≦0.01[3]	1.08±0.37 n.s.[3]
	0.96±0.07 (22)	1.20±0.09 ≦0.02		0.96±0.11 (18)[4]	1.27±0.08 ≦0.01	
$\frac{\text{Urinary lysozyme}}{\text{Urinary creatinine}}$, μg/mg	0.64±0.43 (10)	0.90±1.01 n.s.[2]	0.78±0.99 n.s.[2]	0.71±0.60	0.57±0.26 n.s.[3]	0.35±0.43 n.s.[3]
	1.32±0.40 (17)	0.78±0.22 n.s.[2]		0.73±0.61 (13)[4]	1.06±1.02 n.s.[3]	

Numbers in parentheses represent number of patients. Two sets of numerical data are presented as mean±SEM for urinary α-glucosidase and lysozyme excretion, the upper representing patients followed sequentially until day 10, and the lower those patients in whom data were obtained only before and after MTX and 5-FU administration.

[1] Cytoxan (CTX) 100 mg/m²/day, orally. Methotrexate (MTX) 40 mg/m² i.v. on day 0 and day 7. 5-Fluorouracil (5-FU) 600 mg/m² i.v. on day 0 and day 7.

[2] Wilcoxon's test for paired data vs basal values. Numbers refer to level of significance.

[3] Wilcoxon's test for paired data vs day 7 values. Numbers refer to level of significance.

[4] The second dose of MTX and 5-FU was not administered to 4 patients.

urinary lysozyme increase partly derives from leukocyte lysis. Further studies are needed to detect any difference in nephrotoxicity of various cytostatic agents, chiefly in patients with different degrees of renal function impairment.

These preliminary studies seem to suggest that urinary enzymes are a very sensitive index of tubular damage, and a simple and useful method for the evaluation of drug nephrotoxicity.

Summary

Urinary α-glucosidase, lysozyme, β_2-microglobulin and the electrophoretic pattern of urinary proteins were determined after intravenous administration of potentially nephrotoxic agents (radiological contrast media, cytostatic drugs).

Tubular damage, generally slight and reversible, was suggested by the increase of enzymuria.

Urinary enzymes may be regarded as a sensitive index of tubular damage, and their determination seems worth being performed in high-risk patients.

Acknowledgements

A preliminary report of this work was presented at the XVIII Congress of the Italian Society of Nephrology, Ancona, 1977.

The authors wish to thank Miss Fedora Tamburini for secretarial assistance.

References

1 Bengtsson, U.; Cederbom, G.; Falkheden, T., and Jagenburg, R.: Clearances of inulin and para-aminohippurate before and after urography using high dosage of contrast medium in patients with renal insufficiency. Scand. J. Urol. Nephrol. *2:* 173–181 (1968).

2 Bonini, P.A. and Ceriotti, G.: Human urinary α-glucosidase as an index of kidney tubular damage. Clin. chim. Acta *27:* 415–419 (1970).

3 Bonini, P.A.; Guarnieri, G.F.; Nadai-Frank, A. de e Ceriotti, G.: Significato fisiopatologico e applicazioni cliniche degli enzimi urinari. Quad. Sclavo Diagn. *9:* 419–441 (1973).

4 Burchardt, U; Peters, J.E.; Neef, L.; Thulin, H.; Gründig, C.A. und Haschen, R.J.: Der diagnostische Wert von Enzymbestimmungen in Harn. Z. Med. Labor.-Diagn. *18:* 190–212 (1977).

5 Campanacci, L.; Guarnieri, G.F.; Faccini, L.; Bellini, G., and Carretta, R.: Utilità di alcune indagini enzimologiche in nefrologia clinica. Min. Nefrol. *22:* 155–169 (1975).

6 CERIOTTI, G. and GUARNIERI, G.F.: Physiological background for a possible utilization of urinary enzymes in clinical diagnosis; in BURLINA Proc. 6th Int. Symp. on Clin. Enz., pp. 255–270 (Kurtis, Milano 1974).

7 CERIOTTI, G. et GUARNIERI, G.F.: L'interesse diagnostico della determinazione degli enzimi urinari. Quad. Sclavo Diagn. *11:* 103–121 (1975).

8 COSTANTINI, A. e PALMA, L. DALLA: Urografia mediante fleboclisi rapida. Boll. Soc. Urol. Centro-Merid. *3:* 1–145 (1967).

9 DAVIDSON, A.J.: Lysozymuric response to contrast material in the dog. Investve Radiol. *3:* 65–71 (1968).

10 DUBACH, U.C.: Enzymes in urine and kidney. Proc. Int. Conf. on Enzymes in Urine, Rheinfelden 1967 (Huber, Berne 1968).

11 GRUSKIN, A.B.; OETLIKER, O.H., and WOLFISH, N.M.: Effects of angiography on renal functioning and histology in infants and piglets. J. Pediat. *76:* 41–48 (1970).

12 GOLDSTEIN, E.J.; FEINFELD, D.A.; FLEISCHNER, G.M., and ELKIN, M.: Enzymatic evidence of renal tubular damage following renal angiography. Radiology *12:* 617–619 (1976).

13 GUARNIERI, G.F.; FACCINI, L.; BELTRAM, E.; CHERSICLA, M.; BATTILANA, G.; BRANCHINI, B.; FERRARA, P.; MICHELI, W., and CAMPANACCI, L.: Urinary enzymes (α-glucosidase, lysozyme), protein pattern and β_2-microglobulin as indexes of tubular damage. I. After cardiac surgery (this volume).

14 GUARNIERI, G.F.; FACCINI, L.; CHERSICLA, M.; BELTRAM, E.; BATTILANA, G.; MILANI, S.; LIN, S. e CAMPANACCI, L.: Indagini enzimologiche sull'integrità del tubulo renale in corso di polichemioterapia antiblastica, dopo somministrazione e.v. di mezzi di contrasto ed in operati in circolazione extracorporea. Atti XVIII Congr. Soc. It. Nefrol. Ancona 1977 (in press).

15 HARRISON, J.F.; LUNT, G.S.; SCOTT, P., and BLAINEY, J.D.: Urinary lysozyme, ribonuclease, and low-molecular-weight proteins in renal disease. Lancet *i:* 371–375 (1968).

16 KLEINKNECHT, D.; JUNGUS, P. et MICHEL, J.R.: Les accidents de néphrose osmotique après urographie par perfusion chèz l'insuffisant renal. Ann. Med. intern. *122:* 1167–1171 (1971).

17 MILMAN, N. and STAGE, P.: High-dose urography in advanced renal failure. II. Influence on renal and hepatic function. Acta radiol. *15:* 104–112 (1974).

18 MILMAN, N. and GOTTLIEB, P.: Renal function after high-dose urography in patients with chronic renal insufficiency. Clin. Nephrol. *7:* 250–254 (1977).

19 OLDER, R.A.; MILLER, J.P.; JACKSON, D.C.; JOHNSRUDE, I.S., and THOMPSON, W.M.: Angiographically induced renal failure and its radiographic detection. Am. J. Roentg. *126:* 1039–1045 (1976).

20 PROCKOP, D.J. and DAVIDSON, W.S.: A study of urinary and serum lysozyme in patients with renal disease. New Engl. J. Med. *270:* 269–274 (1964).

21 RAAB, W.P.: Diagnostic value of urinary enzyme determinations. Clin. Chem. *18:* 5–25 (1972).

22 TRNAVSKY, K. and ROVENSKY, J.: Influence of some antirheumatic drugs and cytostatics on urinary enzyme level. Pharmacology *14:* 378–384 (1976).

G.F. GUARNIERI, MD, Istituto di Patologia Medica dell'Università di Trieste, c/o Ospedale Maggiore, Piazza Ospedale 1, *I-34129 Trieste* (Italy)

Enzymes in Health and Disease. Inaug. Scient. Meet. Int. Soc. Clin. Enzymol., London 1977, pp. 201–205 (Karger, Basel 1978)

Urinary Excretion of Trehalase, *N*-Acetyl-β-Glucosaminidase and β-Glucuronidase in Patients with Glomerulopathies

D. Maruhn, H. Hartmann, D. Heredia, L. Singchoovong and K. D. Bock

Division of Clinical Chemistry and Division of Renal and Hypertensive Disease, University (GHS) of Essen, Essen

Introduction

Increased output of some urinary enzymes in glomerular kidney disease is known [8, 9], but the origin and diagnostic relevance of these findings is still under discussion [2]. We have studied whether the determination of a rather kidney-specific enzyme (trehalase, TRE, EC 3.2.1.28) and two enzymes of lysosomal origin (*N*-acetyl-β-glucosaminidase, AGS, EC 3.2.1.30, and β-glucuronidase, GRS, EC 3.2.1.31) in urine may contribute to the detection of glomerulopathies and whether typical patterns are associated with different types of disease.

Material and Methods

Patients. 21 female and 29 male patients (age 14–60 years) with predominantly glomerular kidney diseases. 46 suffered from biopsy-proved glomerulonephritis, 3 subjects had amyloidosis and one myeloma kidney. 13 of these patients had nephrotic syndrome. Patients were divided into four subgroups according to renal function and those with diffuse glomerulonephritis also according to morphological criteria established by Bohle *et al.* [1].

Group I. 29 patients with normal renal function (creatinine clearance >95 ml/min and 1.73 m^2 body surface).

Group II. 21 patients with impaired renal function (creatinine clearance: 6–85 ml/min and 1.73 m^2 body surface).

Group III. 31 subjects with mesangioproliferative forms of glomerulonephritis.

Group IV. 15 patients with focal sclerosing, perimembranous, and membranoproliferative glomerulonephritis. Groups III and IV did not differ in respect to renal function.

Controls. 100 healthy subjects served as controls [5].

Urine was collected between 6 and 9 a.m. and subjected to gel filtration before enzyme assay [10]. AGS and GRS were measured photometrically with the appropriate 4-nitrophenylglycoside [5, 6]. TRE was determined by an adaption of the method of COURTOIS *et al.* [3]. Enzyme activities were calculated as U/l (37 °C), enzyme *output* was expressed as mU/3 h, corrected for average adult body surface, 1.73 m^2.

Protein was measured by a modification of the method of PATRICK and THIERS [7]. Non-parametric tests were used for the statistical analysis.

Results

Normal limits for the urinary excretion of the three enzymes were (mU/3 h): TRE, females 16–132, males 12–148; GRS, females 40–358, males 68–437; AGS, females 309–1,575, males 261–1,825 [5].

Raised urinary levels of trehalase occurred in 21 of 50 patients (42%), of β-glucuronidase in 25 of 50 (50%), and of *N*-acetyl-β-glucosaminidase in all but one patient (98%). Protein excretion was increased in 43 patients (86%)

A significantly elevated output of TRE, AGS and GRS was observed in patients with glomerulopathies and normal renal function. Patients with impaired kidney function (group II) showed normal average trehalase and a moderately but significantly increased output of GRS. The average excretion of AGS was similar to that of patients without azotaemia. A comparison of enzyme excretion between both groups is shown in figure 1.

In spite of these differences, the excretion of the enzymes was not correlated to creatinine clearance *within* these subgroups (Spearman rank correlation coefficients r_s ranging from –0.26 to +0.25).

Classification of patients with glomerulonephritis according to morphological criteria revealed obvious differences in the urinary enzyme pattern. Patients of group IV had a significantly higher urinary output of TRE and AGS than subjects of group III (fig. 2). The difference in average GRS excretion (median value group III: 386 mU/3 h, group IV: 705 mU/3 h) was not statistically significant. Average protein excretion was higher in group IV (median: 626 mg/3 h) compared to group III (median: 231 mg/3 h). Only 6 of 31 patients of group III were found to have pathological results for all 3 enzymes, in contrast to 9 of 15 patients in group IV. 10 of the 13 patients with nephrotic syndrome excreted raised quantities of all 3 enzymes.

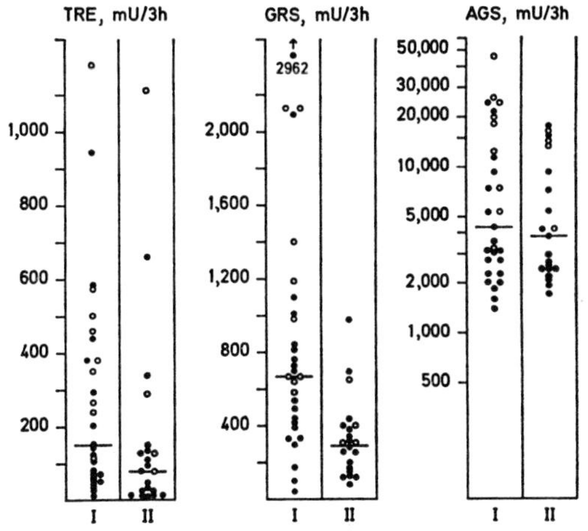

Fig. 1. Urinary enzyme pattern in patients with glomerulopathies without (I) and with (II) impaired renal function. ○ = Nephrotic syndrome. Horizontal lines: median values.

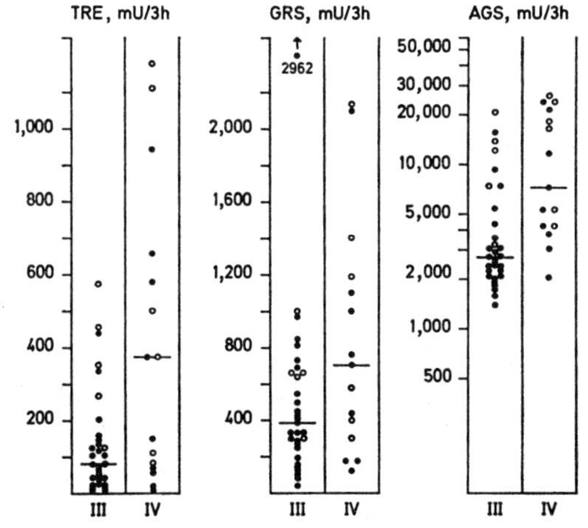

Fig. 2. Urinary enzyme pattern in different morphological types of glomerulonephritis (groups III and IV, see Material and Methods). ○ = Nephrotic syndrome. Horizontal lines: median values.

A correlation was found between TRE and AGS excretion and the degree of simultaneously measured proteinuria (rank correlation coefficients ranging from 0.48 to 0.59). The output of GRS was independent of the degree of proteinuria.

Discussion

The determination of urinary enzymes is of known value in the detection of acute inflammatory and toxic renal damage. In chronic glomerular kidney disease, enzyme output has been related either to the activity of the disease or to the deterioration of renal function [4, 9]. The results of this study confirm that the determination of *N*-acetyl-β-glucosaminidase may be helpful in the detection of chronic glomerulopathies. In all but one patient, an increased output was found and peak activities were about 25 times the upper normal limit. The one patient with an AGS value near the upper normal limit had a minimal proliferating intercapillary glomerulonephritis. The excretion of AGS seems to be independent of kidney function but was correlated to urinary protein output. This does not necessarily reflect increased glomerular filtration of this enzyme. In view of its molecular weight (about 160,000) and its localization in the kidney, it seems possible that AGS is released from tubular lysosomes by exocytosis due to excessive protein reabsorption.

In contrast to AGS, urinary GRS did not appear to be equally suited for diagnostic purposes. Only 50% of the patients revealed an elevated output. In patients with chronic renal insufficiency, GRS was lower than in patients with normal renal function. In this respect, the output pattern of GRS resembles more trehalase, a membrane-bound enzyme occurring in the kidney exclusively in proximal tubular cells. The occurrence of increased amounts of this rather kidney-specific enzyme in 42% of the patients (predominantly in those with normal renal function) may reflect raised glomerular filtration and/or additional injury to tubular brush border.

The reason for the different TRE and AGS excretion in morphologically distinct forms of glomerulonephritis is unknown, but this finding may be attributed to the different degrees of proteinuria. Since there is a wide overlapping of values, no aid in differential diagnosis of glomerulonephritis can be expected from this urinary enzyme pattern at the present time.

We conclude that the determination of AGS offers a useful contribution to the recognition of glomerular kidney diseases. Especially high values occur in heavy proteinuria, but diagnostic sensitivity of the enzyme test seems to be superior to the determination of protein excretion. In 6 patients with chronic glomerulonephritis, a normal protein output but raised excretion of *N*-acetyl-β-glucosaminidase was found.

References

1 Bohle, A.; Eichenseher, N.; Fischbach, H.; Neild, G.H.; Wehner, H.; Edel, H.H.; Losse, H.; Renner, E.; Reichel, W., and Schütterle, G.: The different forms of glomerulonephritis. Morphological and clinical aspects, analyzed in 2500 patients. Klin. Wschr. *54:* 59–73 (1976).

2 Burchardt, U.; Peters, J.E.; Neef, L.; Thulin, H.; Gründig, C.A. und Haschen, R.J.: Der diagnostische Wert von Enzymbestimmungen im Harn. Z. med. Labor.-Diagn. *18:* 190–212 (1977).

3 Courtois, J.E.; Demelier, J.F.; Labat, J. et Bark, C.: Recherches sur la tréhalase urinaire humaine. C. r. Séanc. Soc. Biol. *10:* 2092–2094 (1969).

4 Dance, N.; Price, R.G.; Cattell, W.R.; Lansdell, J., and Richards, B.: The excretion of *N*-acetyl-β-glucosaminidase and β-galactosidase by patients with renal disease. Clin. chim. Acta *27:* 87–92 (1970).

5 Maruhn, D.; Fuchs, I.; Mues, G., and Bock, K.D.: Normal limits of urinary excretion of eleven enzymes. Clin. Chem. *22:* 1567–1574 (1976).

6 Maruhn, D.: Rapid colorimetric assay of β-galactosidase and *N*-acetyl-β-glucosaminidase in human urine. Clin. chim. Acta *73:* 453–461 (1976).

7 Patrick, R.L. and Thiers, R.E.: The direct spectrophotometric determination of protein in cerebrospinal fluid. Clin. Chem. *9:* 283–295 (1963).

8 Szasz, G.; Baranyai, P.; Czirbesz, Z. und Csaki, P.: Über die Enzymurie bei nephrotischem Syndrom. Klin. Wschr. *43:* 783–788 (1965).

9 Wellwood, J.M.; Ellis, B.G.; Price, R.G.; Hammond, K.; Thompson, A.E., and Jones, N.F.: Urinary *N*-acetyl-β-D-glucosaminidase activities in patients with renal disease. Br. med. J. *iii:* 408–411 (1975).

10 Werner, M.; Maruhn, D., and Atoba, M.: Use of gel filtration in the assay of urinary enzymes. J. Chromat. *40:* 254–263 (1969).

D. Maruhn, MD, Zentrales Klinisch-Chemisches Laboratorium, Medizinische Klinik und Poliklinik der Universität (GHS) Essen, Hufelandstrasse 55, *D-4300 Essen 1* (FRG)

Enzymes in Health and Disease. Inaug. Scient. Meet. Int. Soc. Clin. Enzymol., London 1977, pp. 206–208 (Karger, Basel 1978)

The Importance of the Determination of 2,3-Diphosphoglycerate (DPG) in Persons with Respiratory Insufficiency

The Relation of DPG to Age in Normal Persons

E. De la Morena, L. A. Garcia-Lomas and J. Moneo

Biochemistry Department, Fundación Jiménez Díaz, School of Medicine, Universidad Autónoma, Madrid

In recent years, it has been demonstrated that DPG is the phosphorylated compound present at the highest concentration in human erythrocytes; its influence on the dissociation curve of hemoglobin and its molar relation to hemoglobin have also been demonstrated [1, 2, 4, 5]. These physiological facts have led us to study a series of individuals with pathological processes of the respiratory tract. At the same time, we have studied the concentration of DPG in relation to age in normal persons.

Material and Method

The disease processes of the patients had been previously diagnosed clinically and radiologically, and blood samples were taken before treatment (chemotherapy or radiotherapy). The normals were blood donors or persons who had had check-ups without showing any pathology, and all were males in order to avoid possible sex-related influences (both pathological and normal). The blood was taken without stasis from the basilic vein, and the individuals had been fasting for 12 h previously.

The method used to determine DPG was the ultraviolet combination test of the Boehringer Mannheim firm used by Ericson and de Verdier [6].

Results

As illustrated in table I, all processes that involve respiratory insufficiency show a considerable increase of the concentration of DPG in blood. Cases of cancer of the lung show the greatest increase, cases of chronic bronchitis

Table I. DPG: μmol/ml whole blood

	Number	Mean (range)	SD
Cancer of the lung	34	4.61 (6.45–4.13)	0.61
Chronic bronchitis	26	3.77 (4.16–3.55)	0.14
Acute bronchitis	32	3.09 (3.35–2.72)	0.21
Pulmonary tuberculosis	6	3.84 (4.39–3.35)	0.33
Acute pneumonitis	21	2.89 (3.35–2.71)	0.13

Table II. DPG: μmol/ml whole blood

Age, years	Number	Mean (range)	SD
21–30	100	1.97 (2.80–1.55)	0.33
31–40	100	2.11 (2.80–1.85)	0.26
41–50	100	2.15 (2.30–1.90)	0.10
51–60	73	2.25 (2.30–2.15)	0.05
61–70	55	2.61 (2.90–2.35)	0.22
71–80	20	3.04 (3.25–2.90)	0.13

and pulmonary tuberculosis showed high concentrations of DPG, and acute pneumonic and bronchitic processes showed lesser amounts of DPG in the blood although the figure is still higher than normal.

Our results for normal persons coincided with those reported by other investigators [7]: 1.6–2.6 μmol/ml of whole blood.

We found (table II) that DPG increases with increasing age. Although the increase from one decade to the next is small, if we compare the concentration in the early years with that of the last group (71–80 years of age), we see that the increase of DPG is quite large.

Discussion

Human erythrocytes contain concentrations of DPG approximately equal to the concentration of hemoglobin, and their molecules of hemoglobin are arranged in such a way that the positively charged chains of the β-subunits are in position to react with the negative charges of DPG. However, this can occur only when the molecule is in a straight configuration, with the β-subunits separated. Certainly, the combination with DPG makes it necessary for the subunits to adapt to a configuration in which the combination with oxygen is less favorable. It seems that DPG must be expelled from the molecule before it can combine with the oxygen. The greater the concentration of DPG the greater must be the concentration of DPG in the erythrocyte, the smaller the proportion of oxygen that may be retained at a given partial pressure. Since the resulting loss of affinity has little effect on the saturation of the hemoglobin by the oxygen in the lungs, the increase of DPG mainly stimulates the loss of oxygen in the tissues. Changes of altitude [3], diseases with respiratory insufficiency due to pulmonary lesions or other conditions that affect the supply of oxygen (smoking or age) cause adjustments in the concentration of DPG in the erythrocytes as a mechanism of adaptation of the organism to the new physiopathological situation.

References

1 Benesch, R. and Benesch, R.E.: Effect of organic phosphates from human erythrocytes on the allosteric properties of haemoglobin. Biochem. biophys. Res. Commun. *26:* 162–167 (1967).

2 Chanutin, A. and Curnisch, R.: Effect of organic and inorganic phosphates on the oxygen equilibrium of human erythrocytes. Archs biochem. Biophys. *121:* 96–102 (1967).

3 Lenfant, C.; Torrance, J.; English, E.; Finch, C.A.; Reynafarje, C.; Ramos, J., and Faura, J.: Effect of altitude on oxygen binding by hemoglobin and on organic phosphate levels. J. clin. Invest. *47:* 2652–2656 (1968).

4 Luque, J.; Diederich, D., and Grisolia, S.: Binding of 2,3-diphosphoglycerate to oxyhemoglobin. Biochem. biophys. Res. Commun. *36:* 1019–1023 (1969).

5 De la Morena, E.; García-Lomas, L.A.; Latour, P., and Moneo, F.J.: Influence of tonometry on the values of 2,3-diphosphoglycerate (DPG) in normal blood. Proc. Int. Physiological Sciences, vol. XIII, p. 502 (1977).

6 Ericson, A. and Verdier, C.H. de: A modified method for the determination of 2,3-diphosphoglycerate in erythrocytes. Scand. J. clin. Lab. Invest. *29:* 85–90 (1972).

7 Grisolia, S.; Moore, K.; Luque, J., and Grady, H.: Automatic procedure for the microestimation of 2,3-diphosphoglycerate. Analyt. Biochem. *31:* 235–245 (1969).

Prof. E. De la Morena, Capitán Haya 20, *Madrid 20* (Spain)

Enzymes in Health and Disease. Inaug. Scient. Meet. Int. Soc. Clin. Enzymol., London 1977, pp. 209–212 (Karger, Basel 1978)

The Measurement of Serum UDP Galactosyl Transferase in Diabetic Patients

C. P. Price, J. Hutton and K. G. M. M. Alberti

Department of Chemical Pathology and Human Metabolism, Southampton General Hospital, Southampton

Introduction

In recent years, it has been accepted that the vascular complications of diabetes involve thickening of basement membranes [4]. This change in basement membrane has been observed in kidney, muscle, nerve and retina and, in experimental animals, has been related to the duration of diabetes. Changes may be associated with factors not directly related to the organ involved, as thickening has been observed in previously normal kidneys transplanted into diabetic patients.

Studies of the enzymes involved in basement membrane synthesis have shown an increase in UDP glucosyl transferase in the kidney cortex in diabetic animals. This increase returns to normal in short-term diabetic animals treated with insulin. Increases in UDP galactosyl transferase were also observed in diabetic kidney cortex [3].

Knowledge of the pathogenesis of diabetic microangiopathy is essential to provide a basis for rational treatment and prevention of the condition.

Materials and Methods

Blood samples were collected from 47 diabetic patients of both sexes (ages 16–75 years). Samples were also obtained from 56 control subjects of both sexes (ages 19–72 years).

The serum UDP galactosyl transferase activity was measured by determining the amount of galactose transferred to desialised, degalactosylated fetuin (DSG fetuin) [2]. The assay mixture contained MES buffer 12.5 μmol, pH 6.8, UDP galactose [^{14}C] 5 nmol, 0.01 μCi, DSG fetuin 0.75 mg, $MnCl_2$ 0.625 μmol, in a volume of 50 μl. Sample (5 μl) was incubated at 37 °C for 1 h and the reaction terminated by the addition of 1,000 μl of 5%

trichloroacetic acid in 1% phosphotungstic acid. The precipitate was washed three times in precipitant, suspended in 500 μl of saline and then transferred to a scintillation vial with 4 ml of scintillation fluid. The within batch precision was 4.3% (CV) and between batch 7.3% (CV).

Random blood glucose was determined by a hexokinase procedure. Serum creatinine was determined using the Jaffe reaction on a Vickers M300 analyser. Assessment of urine glucose was by 'Labstix' (Ames Company, Stoke Poges, UK). Retinopathy was assessed by two ophthalmologists independently, from colour photographs taken around and including the optic disc. Retinal changes were graded against standard retinal photographs.

Results

The serum UDP galactosyl transferase level in the control subjects gave a mean value of 82.6 μmol/l/h (SD 10.4) with a range of 60–106; there was no correlation with age or sex. Of the 47 diabetic patients, 40 were on insulin therapy; no attempt has been made to identify differences between insulin treatment and oral hypoglycaemic drug therapy. The serum UDP galactosyl transferase activity was elevated in 15 of the 47 patients. An unpaired *t*-test

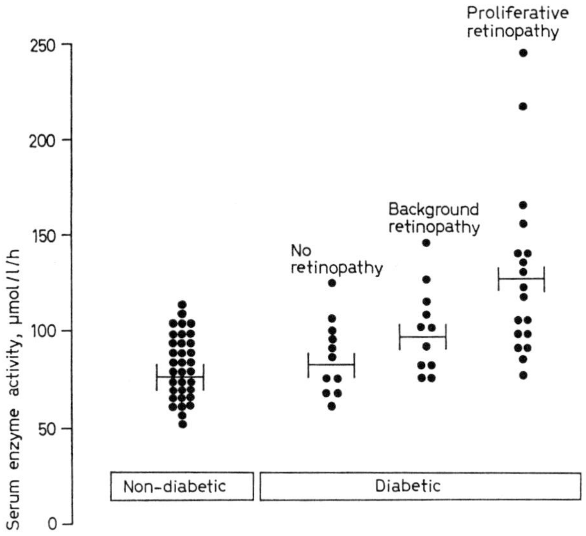

Fig. 1. Serum UDP galactosyl transferase levels in diabetics in relation to retinopathy. Bars represent mean activity in each group.

on log-transformed data showed the activity in the diabetic group to be significantly increased compared with the control group ($p<0.005$).

Retinal changes were graded as absent, background retinopathy and proliferative retinopathy. Comparison of the enzyme activity with these changes (fig. 1) indicated that levels in patients with proliferative retinopathy were increased compared with controls and diabetics with minimal changes. Despite these results, some patients with proliferative retinopathy had normal enzyme levels.

Comparison of enzyme levels with serum creatinine showed a significant correlation ($r=0.75$, $p<0.001$). However, comparison of enzyme levels with random blood glucose did not show a significant correlation, despite the fact that those patients with a random blood glucose of greater than 10 mmol/l had higher enzyme activities than those with glucose levels below 10 mmol/l.

Assessment of the degree of diabetic control was measured by the mean urinary glucose over a 2-week period and showed that in patients with relatively poor control there were significantly higher serum enzyme levels (fig. 2).

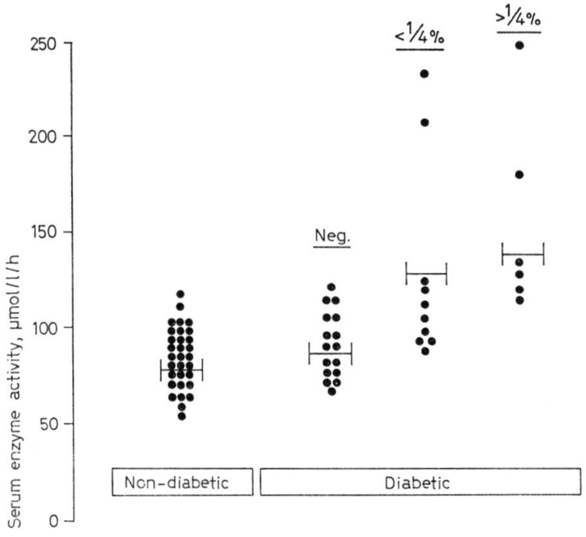

Fig. 2. Serum UDP galactosyl transferase levels in diabetics in relation to the degree of diabetic control assessed by mean urine glucose over a 2-week period. Bars represent mean activity in each group.

Discussion

Alteration in the morphology of basement membrane of the kidney and retina is well recognised. These changes are thought to be the result of accelerated synthesis of basement membrane; however, changes in other tissues have also been observed. The results presented in this paper show a significant correlation between alteration in renal function and serum UDP galactosyl transferase. Furthermore, there is a significant increase in the serum enzyme level in diabetics with proliferative retinopathy. Whilst little is known of the interrelationships of the various vascular complications, it is clear that further objective comparison is not feasible as the enzyme changes may reflect multiorgan involvement.

It has been suggested that changes related to specific glycosylation reactions may contribute to the vascular complications of diabetes; glycosylated haemoglobins have been shown to reflect the degree of diabetic control [1]. The increase in serum UDP galactosyl transferase in patients with significant glycosuria suggests that the level of enzyme may reflect the degree of diabetic control. If these observations are confirmed in long-term studies, the measurement of serum UDP galactosyl transferase may provide a more objective assessment of diabetic control.

Summary

Serum UDP galactosyl transferase was measured in 47 diabetics and 56 control subjects. The enzyme was elevated in patients with proliferative retinopathy and showed significant correlation with serum creatinine. The enzyme was elevated in patients who were poorly controlled, as assessed by urinary glucose measurement over a 2-week period.

References

1 KOENIG, R.G.; PETERSON, C.M.; JONES, R.L.; SAUDEK, C.; LEHRMAN, M., and CERAMI, A.: Correlation of glucose regulation and haemoglobin A_{1c} in diabetes mellitus. New Engl. J. Med. *295:* 417–420 (1976).

2 SPIRO, R.G.: Periodate oxidation of the glycoprotein fetuin. J. biol. Chem. *239:* 567–573 (1964).

3 SPIRO, R.G.: Biochemistry of the renal glomerular basement membrane and its alterations in diabetes mellitus. New Engl. J. Med. *288:* 1337–1342 (1973).

4 WILLIAMSON, J.R. and KILO, C.: Current status of capillary basement membrane disease in diabetes mellitus. Diabetes *26:* 65–75 (1977).

C.P. PRICE, PhD, Southampton General Hospital, Department of Chemical Pathology and Human Metabolism, *Southampton* (England)

Enzymes in Health and Disease. Inaug. Scient. Meet. Int. Soc. Clin. Enzymol., London 1977, pp. 213–217 (Karger, Basel 1978)

Diagnostic Value of Lysozyme Activity Estimations in Pathological Body Fluids

A. PRUSS, Z. SAWICKI and J. KRAWCZYNSKI

Department of Laboratory Diagnostics, Medical Centre of Postgraduate Education, Warszawa

Introduction

The activity of lysozyme has rarely been estimated in pathological pleural and peritoneal fluids [2, 6]. The diagnostic value of such estimations is as yet virtually unknown. We have undertaken the investigation with the aim of evaluating the diagnostic value of lysozyme (LZM) activity estimations in pathological body fluids in certain diseases in the course of which pathological fluid accumulates in the pleural or peritoneal cavity.

Material and Methods

The clinical material consisted of 101 patients with pathological fluid in the pleural or peritoneal cavity, who were classified as follows: (a) pulmonary tuberculosis (26 patients); (b) neoplastic diseases (41 patients), and (c) non-specific inflammatory diseases (exudative non-specific pleurisy, traumatic pleurisy, empyema, peritonitis after surgery or after intestinal inflammation) (34 patients).

Control group I consisted of 90 healthy persons and control group II of 5 patients with circulatory insufficiency and liver cirrhosis with transudates in the pleural or peritoneal cavity.

LZM activity in blood serum and in fluid was estimated by a modified turbidimetric method [3].

The results were expressed as LZM activity in the fluid (LZM_{fl}) and in the serum (LZM_s) and as 'cellular fraction of LZM activity' in the fluid (LZM_{cf}) calculated from the formula:

$$LZM_{cf} = LZM_{fl} - (a + b \cdot LZM_s),$$

where a and b are constants estimated experimentally as 0.282 and 0.731.

Table I. LZM activity in pathological body fluids (LA_{fl}), 'cellular fraction' of LZM activity in fluids (LA_{cf}) and LZM activity in the serum (LA_s) in control groups and in patients with diseases characterized by accumulation of pathological fluid in pleural and/or peritoneal cavity (in U/ml, mean values and range)

Parameter investigated	1st control group	2nd control group	Patients with pathological changes of serous membranes with: tuberculosis	neoplastic disease	non-specific inflammatory diseases
	(n=90)	(n=5)	(n=26)	(n=41)	(n=34)
LA_{fl}	–	6.8	16.7(19)[a]	8.28(6)[a]	24.3(17)[a]
		2.0–11.6[b]	4.5–34.9	2.4–29.3	3.5–125.4
LA_{cf}	–	–0.9	10.33(26)[a]	0.5(25)[a]	15.7(21)[a]
		–2.3– 0.5[b]	1.9–25.8	–10.4–6.5	6.7–111.4
LA_s	7.4	6.8	8.6(5)[a]	10.3(8)[a]	11.5(12)[a]
	3.0–11.8[b]	4.76–8.87	1.4–15.6	3.3–38.9	2.8–31.5

[a] Number of cases with elevated values.
[b] Mean values ±2 standard deviations.

Results

Reference values. The average activity of LZM in blood serum (LZM_S) in control group I was 7.4±4.4 U/ml (range: 3.0–11.8) the average LZM activity in transudate fluid: 6.8 U/ml (range: 2.0–11.6) and the mean values of 'cellular fraction' (LZM_{cf}): –0.9±1.4 U/ml (range: –2.3 to –0.5) in control group II.

LZM activity in patients. The average values obtained in controls and patients are summarized in table I.

(a) Pulmonary tuberculosis: The activity of LZM in pleural fluid (LZM_{fl}) was elevated in 19 cases (all exudates). The values of LZM_{cf} were considerably elevated in all 26 patients. LZM_s was elevated only in 5 cases. The lowest values of all parameters were found in 2 transudates (LZM_{fl}: 5.3 and 4.5 U; LZM_{cf}: –2.2 and 4.0 U; LZM_s: 1.4 and 2.8 U).

(b) Neoplastic diseases: LZM_{fl} was elevated only in 6 cases. The LZM_{cf} values were well differentiated (negative, zero, positive) being elevated in 25 fluids of which 4 were transudates. LZM_s was elevated only in 8 cases. Two LZM_s values were very high (30.4 and 38.9 U).

(c) Non-specific inflammatory diseases: In this group, elevated LZM_{fl} was observed in 17 fluids reaching in 6 cases values $>$50 U. LZM_{cf} values were elevated in 21 fluids, in some cases very considerably (in 3 cases above 90 U). LZM_s was elevated only in 12 cases but in 2 exceeded 30 U. The highest LZM_{fl} and LZM_{cf} were obtained in patients with inflammatory complications after surgery and in inflammation of the intestine.

Discussion

The results of our previous investigation in which the LZM was estimated in gastric juice and in faeces [4, 5] suggest that LZM originates predominantly in cellular elements (neutrophilic granulocytes, monocytes and macrophages) known to produce LZM [1, 6] and to be components of inflammatory infiltrations localized in gastric and intestinal mucosa. Therefore, LZM activity estimated in such material could be considered as a measure of intensity of the inflammatory process localized in gastric and intestinal wall.

Quantitative examination of pathological body fluids is difficult owing to lack of a proper reference system. We have chosen as basic reference values: average LZM_s of healthy subjects and average LZM_{fl} in transudates obtained from patients without pathological changes of serous membranes (control group II). From these values, it was possible to calculate that about 73% of LZM is filtered from blood to the transudate through undamaged capillaries.

As in our investigation we deal mostly with exudates, it seems justified to calculate the LZM_{cf} from the formula containing such data as LZM_{fl}, LZM_s and constants reflecting the approximate filtration rate of LZM (see above). In our opinion, LZM_{cf} is a more suitable measure of macrophage-leucocytic cellular response than the ratio: LZM_{fl}/LZM_s used by other authors [7].

Analyzing the results in different group of patients, it must be pointed out:

(a) In patients with pulmonary tuberculosis and with pathological changes in the pleura, the most characteristic feature is that all LZM_{cf} values are elevated. This suggests that cellular elements are the main source of LZM, especially macrophages and less neutrophilic granulocytes, usually present in specific tuberculosis infiltrations in the pleura and the lung.

(b) In patients with neoplastic disease involving serous membranes and neighbouring organs, the most characteristic feature is the marked scatter of the results, especially of LZM_{cf}. This can be explained by variation in the damage to the capillaries by the tumour or its toxins, increased filtration of LZM from blood serum, and by differential activation of cellular elements in connection with immunological processes stimulated by the tumour (increased cellular production of LZM).

(c) In patients with non-specific inflammatory diseases of the pleura or peritoneum, the values of LZM_{fl}, LZM_{cf} and LZM_s were also scattered and in a few cases were very high. The scatter of the results can be explained by the clinical heterogenity of this group as well as by the fact that also in this group both mechanisms were probably involved in accumulation of LZM in pathological fluid.

Highly elevated LZM_{fl} and LZM_{cf} can be due to increased cellular production of LZM (in inflammation of intestine additionally by Paneth cells) in inflammatory infiltrations from which the LZM is diluted sometimes with only a very small amount of fluid, e.g. suppurative peritonitis.

On the basis of our clinical observations, it is possible to consider LZM activity estimation in pathological body fluids only as corroborative evidence of inflammatory activity in a pathological process occurring in serous membranes and neighbouring organs.

Summary

In 101 patients with pathological accumulation of fluid in the pleural or peritoneal cavity, lysozyme (LZM) activities in fluid and in blood serum were estimated. Results were expressed as 'cellular fraction of LZM activity'. LZM appears to be an index of an active inflammatory process in the pleura or peritoneum.

References

1 GORDON, S.; TODD, J., and COHN, Z.: *In vitro* synthesis and secretion of lysozyme by mononuclear phagocytes. J. exp. Med. *139:* 1228–1248 (1974).

2 KLOCKARS, M.; PETTERSON, T.; RISKA, H., and HELLSTRÖM, P.: Pleural fluid lysozyme in tuberculous and nontuberculous pleurisy. Br. med. J. *i:* 1381–1381 (1976).

3 KŁOPOCKI, T.; SAWICKI, Z., and KRAWCZYNSKI, J.: Próba eliminowania aktywatorów i inhibitorów lizozymu z moczu przy pomocy saczenia zelowego. Diagn Lab. *9:* 291–304 (1973).

4 KRAWCZUK, J.; SAWICKI, Z., and KRAWCZYNSKI, J.: Diagnostic value of lysozyme activity estimation in feces of infants with acute diarrhoea. Z. klin. chem. klin. Biochem. *16:* 343–347 (1978).

5 Sadowska, H.; Sawicki, Z., and Krawczynski, J.: Wartośê rozpoznawcza oznaczania aktywności lizozymu soku zoładkowego w toku próby Kaya w róznych chorobach zoładka i dwunastnicy. Diagn. Lab. *11:* 121–134 (1975).
6 Seen, H.; Chu, B.; O'Malley, J., and Holland, J.: Experimental and clinical studies on muramidase (lysozyme). I. Muramidase activity of normal human blood cells and in inflammatory exudates. Acta hemat. *44:* 65–77 (1975).
7 Wardle, E.: Simple method for detection of infection of peritoneum during dialysis. Br. med. J. *ii:* 518–520 (1973).

J. Krawczynski, MD, Department of Laboratory Diagnostics, Central Clinical Hospital, Medical Academy, Banacha Street 1A, *PL-02-097 Warszawa* (Poland)

Enzymes in Health and Disease. Inaug. Scient. Meet. Int. Soc. Clin. Enzymol., London 1977, pp. 218–221 (Karger, Basel 1978)

Erythrocyte Glutathione, Glutathione Reductase and Lactate Dehydrogenase in Some Anaemias

R. J. SPOONER and G. A. McDONALD

Departments of Biochemistry and Haematology, Royal Infirmary, Glasgow

Introduction

The increased levels of serum enzymes found in the anaemias and the rapid fall in activity on successful treatment of megaloblastic anaemia has been known for some time [3]. The activities of red cell enzymes in these conditions has not received as much attention. Early work on erythrocyte glutathione reductase (GR, EC 1.6.4.2) took place before the discovery of the dependence of the enzyme on the riboflavin metabolite FAD, a deficiency of which might easily coexist with several of the anaemias.

The elevation of erythrocyte glutathione in anaemia is well established [5] and work relating this to erythrocyte GR (EGR) in iron deficiency anaemia has been published [6].

The aim of the present study is to consider the interrelationship of glutathione and EGR. Erythrocyte lactate dehydrogenase (LDH, EC 1.1.1.27) was also measured and acted as an internal control of EGR.

Materials and Methods

All patients studied had an initial haemoglobin of less than 10 g% and the sample was obtained before or within 24 h of commencement of treatment. Red cells were washed three times with isotonic saline and then haemolysed in deionised water to give a haemoglobin of approximately 10 g/l.

Enzyme assays were performed in duplicate at 37 °C on an LKB 8206 Reaction Rate Analyser. Glutathione reductase was measured using the method of DELIDES *et al.* [1] incorporating FAD at 5 μm final concentration, LDH using a Boehringer optimised kit, haemoglobin from its absorption spectrum [2] and glutathione was measured fluorimetrically [4].

Table I. Erythrocyte enzyme concentrations and their ratio in patients with anaemia

Group	Number	LDH	GR	LDH/GR
Normal	15	188 ± 27	8.55 ± 1.48	22 ± 4.6
Normochromic/normocytic	6	204 ± 39	10.16 ± 2.74	21 ± 5
Megaloblastic	19	250 ± 66[1]	8.83 ± 3.52	35 ± 13[1]
Hypochromic/microcytic	14	458 ± 143[1]	14.73 ± 2.53[1]	31 ± 9[1]
Haemolytic	3	268	11.29	23

The enzyme concentrations are expressed as U/g Hb ± 1 SD.
[1] Significantly different from the normal $p < 0.01$.

Results

The patients were classified using standard haematological indices, obtained using a Coulter 'S' counter, into megaloblastic, normochromic/normocytic, hypochromic/microcytic and haemolytic anaemias. The megaloblastic group includes 9 patients with pernicious anaemia, 2 with a folate deficiency secondary to adult coeliac disease, 7 with mixed vitamin B_{12} and folate or folate and iron deficiency and 1 patient with a megaloblastic marrow and a hypochromic/microcytic blood film.

Table I gives the enzyme concentrations found and shows a significant elevation of erythrocyte LDH (ELDH) and GR in the hypochromic/microcytic group and a significantly raised ELDH in the megaloblastic group. In both groups, there is an increase in the erythrocyte LDH/GR ratio.

By use of the Kendall rank correlation, a significant association was found between the mean cell volume and each of the erythrocyte enzyme concentrations. For LDH $r = -0.418$, for GR $r = -0.314$ and for both $p < 0.001$.

Erythrocyte glutathione was estimated on a small number of samples and the results confirm its increase in the anaemias.

Serial studies were performed in 10 patients with megaloblastic anaemia. Three patients showed a marked increase in both erythrocyte enzyme concentrations during therapy. Two of these had adult coeliac disease with folate deficiency and the third a mixed deficiency pattern. Figure 1 shows the response obtained in one patient with coeliac disease.

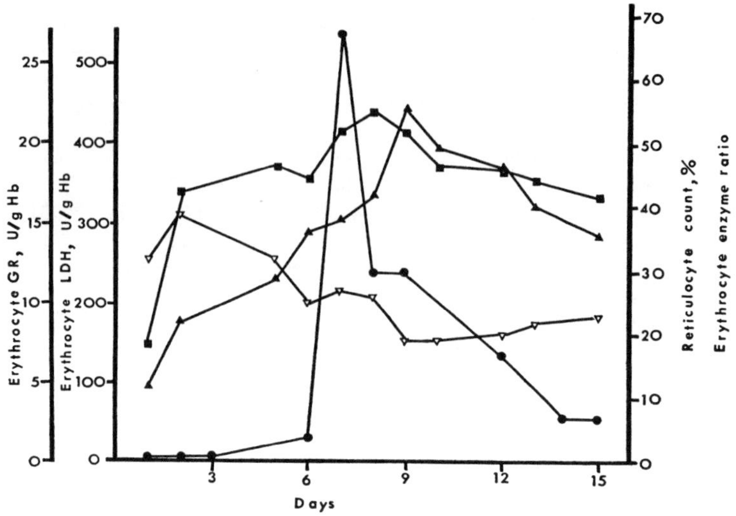

Fig. 1. The response to vitamin $B_{12}$and folic acid therapy of erythrocyte LDH (■), erythrocyte GR (▲), their ratio (▽), and the reticulocyte count (●) in a patient with a megaloblastic anaemia secondary to coeliac disease.

Discussion

The changes observed in erythrocyte enzymes are not due to anaemia *per se* as those in the normochromic/normocytic group are not significant. The regression analyses indicate the changes to be a reflection of the alteration in cell size. The increase in ELDH in the megaloblastic group goes against this theory and the EGR in this group has a very wide scatter (3.75–15.46 U/g Hb).

The mean corpuscular haemoglobin concentration in all these anaemias differs very little from the normal. Because there is a change in intracellular haemoglobin synthesis, this does not necessarily indicate a similar change in synthesis of other erythrocyte proteins.

The sequential studies show that the maximum enzyme response occurs shortly after the maximum reticulocyte count. This reponse to reticulocytosis has been demonstrated using prolonged phlebotomy in sheep [7]. The 3 patients with haemolytic anaemia show a positive relationship between enzyme concentration and reticulocyte count.

The alteration in the ratio of enzymes and its return to normal following reticulocytosis shows a disproportional change in one of the enzyme concentrations. Erythrocyte LDH tends to be elevated on presentation in the megaloblastic group and the greater increase in EGR on reticulocytosis suggests a decay of GR in the mature erythrocyte in this and the hypochromic/microcytic group.

The interrelationship of the erythrocyte enzymes and glutathione is uncertain. It has been shown that glutathione is increased in patients with chronic disease [5], patients usually normochromic/normocytic, a group where we have not shown any significant changes in erythrocyte enzyme concentrations.

Summary

The concentrations of erythrocyte lactate dehydrogenase and glutathione reductase were determined in 57 patients with anaemia. Both enzyme concentrations were significantly raised in the hypochromic/microcytic group and lactate dehydrogenase in megaloblastic patients. A negative correlation was found between each of these concentrations and the mean cell volume. The enzyme ratio was significantly elevated in both these groups. Sequential studies in megaloblastic anaemias show a reduction of this ratio during therapy. A marked increase in enzyme concentrations with a peak around maximum reticulocytosis occurred in 3 patients. Erythrocyte glutathione was shown to be increased in all patients studied.

References

1 Delides, A.; Spooner, R.J.; Goldberg, D.M., and Neal, F.E.: An optimized semiautomatic rate method for serum glutathione reductase activity and its application to patients with malignant disease. J. clin. Path. *29:* 73–77 (1976).

2 Fleming, A.F. and Woolf, A.J.: A spectrophotometric method for the quantitative estimation of bilirubin in Liquor amnii. Clin. chim. Acta *12:* 65–74 (1965).

3 Hess, B.: Diseases of the blood; in Enzymes in blood plasma, pp. 104–108 (Academic Press, New York 1963).

4 Hissin, P.J. and Hilf, R.: A fluorimetric method for determination of oxidised and reduced glutathione in tissues. Analyt. Biochem. *74:* 214–226 (1976).

5 Hopkins, J. and Tudhope, G.R.: Red cell glutathione in anaemia. Scot. med. J. *18:* 177–181 (1973).

6 Ramachandram, M. and Iyer, G.Y.N.: Erythrocyte glutathione reductase in iron deficiency anaemia. Clin. chim. Acta *52:* 225–229 (1974).

7 Smith, J.E. and Lee, M.: How phlebotomy affects the erythrocyte metabolism of high and low glutathione sheep. Res. vet. Sci. *21:* 127–131 (1976).

R.J. Spooner, BSc, Royal Infirmary, Department of Biochemistry, *Glasgow G4 0SF* (Scotland)

Enzymes in Health and Disease. Inaug. Scient. Meet. Int. Soc. Clin. Enzymol., London 1977, pp. 222–227 (Karger, Basel 1978)

Enzymes in Therapy

BRENDA E. RYMAN

Department of Biochemistry, Charing Cross Hospital Medical School, London, and Girton College, Cambridge

The possible use of enzymes for therapeutic purposes has not received much attention in the past, partly due, perhaps, to the attitude of some clinicians who regard enzymes and the enzymologist as an area of pre-clinical medicine and a necessary evil to be overcome in the curriculum of medical training. Currently, however, enzymes are making their presence felt more and more in medicine, and this new era has been aided by clinical enzymology in which the role of serum enzymes in diagnosis [27] has made its mark. For example, transaminase is no longer a term dismissed from the minds of young would-be clinicians on passing from studies of amino acid metabolism to their clinical work.

Enzyme therapy includes the following approaches, some of which are well established and others of which will probably become important during the next decade:

(a) Treatment of pancreatic insufficiency with enzymes in enteric-coated form.

(b) The use of proteolytic enzymes such as trypsin, chymotrypsin and bromelin in inflammatory and traumatic oedema, in the resolution of necrotic tissue, clots and exudates in wounds and cavities, and in the loosening of slough.

(c) The treatment of bites and stings by hyaluronidase. This enzyme has also been suggested (due to the ability to act as a spreading agent) as a possible aid in artificial insemination and in infertility. Furthermore, hyaluronidase is used as a means of treatment of accidental extravasation of irritant drugs and locally in the treatment of painful muscular conditions.

(d) As agents for clots and exudate removal – streptokinase and urokinase are currently finding favour in this context.

(e) The use of asparaginase in the treatment of leukaemias dependent on high levels of asparagine.

(f) While these aspects of enzyme therapy are perhaps the most obvious ones, the use of enzymes to provide essential proteins for therapy of diseases such as kwashiorkor and other states of malnutrition is still in the experimental stage. Thus, efforts are being made to convert readily available world excesses such as glucose and other carbohydrates as well as oil residues to proteins, using bacterial and fungal sources, and obviously such conversions are enzyme-dependent. While such an approach seems immediately applicable to areas of the world where malnutrition exists at the moment, it seems likely that a world shortage of protein will occur in the next 50 years or less that may affect all countries. At an even more sophisticated level, recombinant DNA experiments ('genetic engineering') employing specific enzymes, in which pieces of selected DNA are excised from suitable cells and are spliced into the genetic material of the recipient (usually a bacterium) to form the blue-print for production of some desirable product, must surely find application in therapy, if only at the level of production of insulin for therapeutic purposes by such organisms as *E. coli* following genetic manipulation.

(g) One of the areas which has attracted considerable interest concerns possible enzyme replacement treatment of the genetically inherited disorders in which in many cases a single enzyme is missing, frequently with disastrous effects, e.g. Tay-Sach's disease, Lesch-Nyhan syndrome, Hurler-Hunter mucopolysaccharidoses, and Pompe's glycogenosis. While each such disease is relatively rare, these genetic inborn errors of metabolism make up a group of some 60 well-defined disorders and collectively present a not so rare group [23]. Many of these diseases lead to accumulation of materials resulting from a block in a metabolic pathway due to the lack of a specific enzyme. The injection of purified enzyme directly into the circulation in the hope that the enzyme will be taken up by defective cells has been attempted in glycogen storage disease patients [8, 11, 13, 18] as well as in the lipidoses, particularly Gaucher's and Fabry's diseases [2]. Although in the latter cases some evidence of degradation of accumulated substances in blood, and to a lesser extent in vital organs, occurred, in the case of the glycogen storage diseases results have been equivocal.

Attempted treatment of a patient with a mucopolysaccharidosis with plasma or leucocytes, presumably containing the missing enzyme, has yielded some encouraging results [16], as has such treatment of a patient with Fabry's disease [3, 19]. The use of erythrocytes or erythrocyte ghosts loaded with suitable enzymes has been another approach to introducing the missing protein into vital organs and tissues by IHLER's group [1, 14, 15] and in our laboratory [25]. The possibility of using such erythrocytes in the treatment of Gaucher's disease with its splenic involvement is clearly a fascinating possibility, particularly as the treatment of such erythrocytes with glutaraldehyde increases the spleen targetting [9].

The use of skin in attempted corrective therapy of genetically inherited enzyme deficiency diseases has also received some attention. DEAN *et al.* [7] have reported an increased breakdown of glycosaminoglycans and the appearance of corrective enzymes after skin transplant in a patient with Hurler's disease. Other possibilities of reinstating the enzyme and overcoming the deficiency state include organ transplantation [10].

There are, however, many problems associated with the approaches so far considered and obviously any attempt to inject, infuse or implant enzyme sources are associated with several difficulties, including: (a) target-directing of the enzyme, e.g. if the enzyme deficiency is found in specific organs; (b) possible interaction of therapeutic enzymes with normal tissue elements; (c) triggering of the immune response to foreign protein, and (d) inactivation of the therapeutic enzyme by host proteinase and the relatively short half-life of injected enzyme.

In an attempt to overcome some of these difficulties, attention has been focussed on finding a carrier in which enzymes to be used in replacement therapy could be sequestered. The chemical nature of the selected carriers (called *liposomes)* is phospholipid, which is readily biodegradable and which forms small (25 nm→1 μm diameter), closed vesicle structures into which the enzyme may be entrapped [21]. (For general reviews on liposomes, see COLLEY and RYMAN [6] and TYRRELL *et al.* [24].) Such liposomes are easily formed when dried phospholipids are allowed to swell and become hydrated in the presence of excess aqueous phase. The enzyme activity is protected by the lipid 'overcoat', and after injection of such enzyme-containing liposomes a rapid clearance of liposomes from the circulation is achieved by uptake by liver and spleen. The liposomal contents are found associated with liver lysosomes, and thus a carrier mechanism seems particularly promising in enzyme replacement therapy of lysosomal storage diseases, of which some 30 are known to affect man [17]. While several attempts have been made to test the efficacy of such an enzyme delivery system using liposomes with cells in culture, or in animal model experiments, only recently have we attempted the first therapy in a child with a lysosomal disease [26].

A child with the lethal Pompe's disease (type II glycogen storage disease

– lack of lysosomal α-glucosidase) was treated with liposomes containing amyloglucosidase capable of degrading glycogen [26]. The liposomes, which were made of egg lecithin with cholesterol and phosphatidic acid, appeared to give rise to no toxic effects, and encouraging results were obtained in that the liver size of the child diminished and liver glycogen was lower than was to be expected [26]. Since this first attempted use of liposomes in a patient, they have also been used to treat an adult with Gaucher's disease, also with encouraging results [4].

The problem of tissue directing of enzymes entrapped in liposomes in enzyme replacement therapy still, however, remains unsolved. While entry of such liposomes into the lysosomal apparatus of liver may be useful in Pompe's disease where the liver is enlarged and engorged with glycogen, nevertheless, the muscle accumulation of glycogen (particularly cardiac muscle) is of paramount importance in this disorder and there is little evidence to suggest that liposomes gain access to the body musculature. Likewise, with many other enzyme deficiency disorders, tissues such as brain are involved and liposomes do not appear to cross the blood–brain barrier. Therefore, the use of these phospholipid vesicles in introducing enzymes into defective cells would be of limited value in such conditions. At the present time, efforts are being made to introduce 'homing' agents into the lipid bilayers which make up the liposome structure so that tissue directing may be explored [24]. The attachment of tissue-specific antibodies to the surface of liposomes may aid such an approach. COHEN *et al.* [5] have shown that attachment of human heat-aggregated IgG to liposomes containing entrapped hexosaminidase A leads to their enhanced uptake into leucocytes from Tay-Sach's patients *in vitro*, and GREGORIADIS and NEERUNJUN [12] have shown that liposomes carrying antibody to specific cells are more readily taken up by those cells in culture.

Liposomes have expanded from their earlier role of carriers of enzymes in replacement therapy to that of carriers of drugs, hormones, chelating agents and virus sub-particles, all of which have therapeutic potential [22, 24]. The entrapment of asparaginase in liposomes has recently been shown to be effective in 'treating' mouse leukaemia [20].

Although the possibility of introducing replacement pieces of DNA, containing the purified gene for the missing enzyme, into the coil of mammalian genes in enzyme-deficient diseases is definitely a real one in the future, it will probably be a long time before such strategy can be achieved with specificity and safety. Meanwhile, the possible use of the liposome carrier remains a fascinating challenge and is worthy of further investigation.

References

1 Ang, E.; Glew, R., and Ihler, G.: Enzyme loading of nucleated chicken erythrocytes. Expl Cell Res. *104:* 430–434 (1977).

2 Brady, R.O.; Pentchev, P.G., and Gal, A.E.: Investigations in enzyme replacement therapy in lipid storage diseases. Fed. Proc. Fed. Am. Socs exp. Biol. *34:* 1310–1315 (1975).

3 Brady, R.O.; Tallman, J.F.; Johnson, W.G.; Gal, A.E.; Leahy, W.R.; Quirk, J.M., and Dekaban, A.S.: Replacement therapy for inherited enzyme deficiency. New Engl. J. Med. *289:* 9–14 (1973).

4 Braidman, I. and Gregoriadis, G.: Preparation of glucocerebroside-β-glucosidase for entrapment in liposomes and treatment of patients with adult Gaucher's disease. Biochem. Soc. Trans. *4:* 259–261 (1976).

5 Cohen, C.M.; Weissmann, G.; Hoffstein, S.; Awasthi, Y.C., and Srivastava, S.K.: Introduction of purified hexosaminidase A into Tay-Sachs leukocytes by means of immunoglobulin-coated liposomes. Biochemistry, N.Y. *15:* 452–460 (1976).

6 Colley, C.M. and Ryman, B.E.: The liposome: from membrane model to therapeutic agent. Trends Biochem. Sci. *1:* 203–205 (1976).

7 Dean, M.F.; Muir, H.; Benson, P.F.; Button, L.R.; Batchelor, J.R., and Bewick, M.: Increased breakdown of glycosaminoglycans and appearance of corrective enzyme after skin transplants in Hunter syndrome. Nature, Lond. *257:* 609–612 (1975).

8 Barsy, T. de; Jacquemin, P.; Hoof, F. van, and Hers, H.G.: Enzyme replacement in Pompe disease; in Enzyme therapy in genetic diseases. Birth Defects Original Article series *9:* 184–190 (1973).

9 Deloach, J.; Peters, S.; Pinkard, O.; Glew, R., and Ihler, G.: Effect of glutaraldehyde treatment on enzyme-loaded erythrocytes. Biochim. biophys. Acta *496:* 507–515 (1977).

10 Desnick, R.J.: Allotransplantation in genetic diseases; in Tager, Hooghwinkel and Daems Enzyme therapy in lysosomal storage diseases, pp. 292–297 (North-Holland, Amsterdam 1974).

11 Fernandes, J. and Huijing, F.: Branching enzyme-deficiency glycogenosis: studies in therapy. Archs Dis. Childh. *43:* 347–352 (1968).

12 Gregoriadis, G. and Neerunjun, D.: Homing of liposomes to target cells. Biochem. biophys. Res. Commun. *65:* 537–544 (1975).

13 Hug, G. and Schubert, W.K.: Lysosomes in type II glycogenosis. Changes during administration of extract from *Aspergillus niger*. J. Cell Biol. Commun. C1–C6 (1967).

14 Ihler, G.; Glew, R., and Schnure, F.: Enzyme loading of erythrocytes. Proc. natn. Acad. Sci. USA *70:* 2663–2666 (1973).

15 Ihler, G.; Lantzy, A.; Purpura, J., and Glew, R.H.: Enzymatic degradation of uric acid by uricase-loaded human erythrocytes. J. clin. Invest. *56:* 595–602 (1975).

16 Knudson, A.G.; Ferrante, N. di, and Curtis, J.E.: Effect of leukocyte transfusion in a child with type II mucopolysaccharidosis. Proc. natn. Acad. Sci. USA *68:* 1738–1741 (1971).

17 Kolodny, E.H.: Lysosomal storage diseases. New Engl. J. Med. *294:* 1217–1220 (1976).

18 LAUER, R.M.; MASCARINAS, T.; RACELA, A.S.; DIEHL, A.M., and ILLINGWORTH BROWN, B.: Administration of a mixture of fungal glucosidases to a patient with type II glycogenosis (Pompe's disease). Pediatrics *42:* 672–676 (1968).

19 MAPES, C.A.; ANDERSON, R.L.; SWEELEY, C.C.; DESNICK, R.J., and KRIVIT, W.: Enzyme replacement in Fabry's disease, an inborn error of metabolism. Science *169:* 987–989 (1970).

20 NEERUNJUN, E.D. and GREGORIADIS, G.: Tumour regression with liposome-entrapped asparaginase: some immunological advantages. Biochem. Soc. Trans. *4:* 133–134 (1976).

21 RYMAN, B.E.: Possible use of liposomes in enzyme replacement therapy in storage diseases; in TAGER, HOOGHWINKEL and DAEMS Enzyme therapy in lysosomal storage diseases, pp. 149–162 (North-Holland, Amsterdam 1974).

22 RYMAN, B.E.; JEWKES, R.F.; JEYASINGH, K.; OSBORNE, M.P.; PATEL, H.M.; RICHARDSON, V.J.; TATTERSALL, M.H.N., and TYRRELL, D.A.: Potential application of liposomes to therapy. Ann. N.Y. Acad. Sci. (in press).

23 STANBURY, J.B.; WYNGAARDEN, J.B., and FREDRICKSON, D.S. (eds.): The metabolic basis of inherited disease; 3rd ed. (McGraw-Hill Book Co., New York 1972).

24 TYRRELL, D.A.; HEATH, T.D.; COLLEY, C.M., and RYMAN, B.E.: New aspects of liposomes. Biochim. biophys. Acta *457:* 259–302 (1976).

25 TYRRELL, D.A. and RYMAN, B.E.: The entrapment of therapeutic agents in resealed erythrocyte ghosts and their fate *in vivo*. Biochem. Soc. Trans. *4:* 677–680 (1976).

26 TYRRELL, D.A.; RYMAN, B.E.; KEETON, B.R., and DUBOWITZ, V.: Use of liposomes in treating type II glycogenosis. Br. med. J. *2:* 88–89 (1976).

27 WILKINSON, J.H. (ed.): The principles and practice of diagnostic enzymology (Arnold, London 1976).

B.E. RYMAN, PhD, Department of Biochemistry, Charing Cross Hospital Medical School, Fulham Palace Road, *London W6 8RF* (England)

Enzymes in Health and Disease. Inaug. Scient. Meet. Int. Soc. Clin. Enzymol., London 1977, pp. 228–231 (Karger, Basel 1978)

The Effect of Sodium Bromide on Human Serum Cholinesterase Activity

ANNE D. DICKSON

Biochemistry Department, Royal Infirmary, Glasgow

The studies of KALOW and DAVIES [1953] showed that all the known inhibitors of cholinesterase (acylcholine acyltransferase EC 3.1.1.8) which could differentiate the normal and atypical enzymes contained a positively charged quaternary ammonium group in the molecule, for example, dibucaine. HARRIS and WHITTAKER [1961, 1963] then demonstrated that sodium fluoride and chloride could be of use in differentiating between phenotypes.

As fluoride and chloride are both members of the halogen series, it was decided to investigate the effect of the next number, bromide, on cholinesterase activity.

Experimental

For cholinesterase phenotyping, activity was simultaneously and separately assayed in the presence of 10 mmol/l dibucaine [KALOW and GENEST, 1957], 50 mmol/l sodium fluoride [HARRIS and WHITTAKER, 1961], 500 mmol/l sodium chloride [WHITTAKER, 1968] and 1 mmol/l succinyldicholine chloride [KING and GRIFFIN, 1973].

Assay Procedures

(a) Using a Unicam SP8000 spectrophotometer, cholinesterase activity was estimated by the procedure of KALOW and LINDSAY [1955] modified by KING [1965]. This employs a benzoylcholine substrate and the assay is conducted at 25 °C.

(b) Butyrylthiocholine was used as substrate in a modification of the method of ELLMAN *et al.* [1961]. The assay was carried out at 37 °C on an LKB 8600 reaction rate analyser with LKB 3200 calculator and Data Dynamics 390ASR teletype attached. Both assay procedures were employed and both 67 mmol/l Sörensen phosphate and 200 mmol/l Tris buffers pH 7.4 were used to maintain reaction pH.

The concentrations of bromide used in the investigation ranged from 0.2 to 2 mol/l.

Results

Initial studies with benzoylcholine as substrate showed that sodium bromide did inhibit cholinesterase. The curves obtained were not of the usual sigmoid pattern but showed oscillations. These were thought initially to be caused by changes in pH of the reaction mixture on addition of sodium bromide, but when this was measured it was found that simultaneous oscillations in pH did not occur. It was also noted that there was no marked decrease in inhibition with Tris buffer as might have been expected from the literature [GARRY, 1971].

With the majority of inhibitors in phosphate buffer the usual homozygote $E_1^uE_1^u$ is inhibited most, next $E_1^uE_1^f$, then $E_1^uE_1^a$, $E_1^fE_1^f$, $E_1^aE_1^f$ and finally $E_1^aE_1^a$, the atypical homozygote. This order is reversed with chloride, that is, $E_1^aE_1^a$ is most inhibited. This was also found to occur with bromide.

From the phosphate buffer curves, maximal differentiation of the phenotypes was obtained with a molar concentration of bromide. The inhibition by this concentration of bromide of the six phenotypes was investigated and a comparison of bromide numbers obtained, that is the percentage inhibition, was made with the other inhibitor numbers. It was found that a combination of bromide and dibucaine, or bromide and scoline numbers gave the best differentiation of phenotypes.

The studies with butyrylthiocholine as substrate gave curves of the more usual sigmoid pattern and showed the expected greater inhibitory effect in phosphate than in Tris buffer. The order of inhibition again did not conform to the usual pattern.

Discussion

The bromide studies with benzoylcholine have proved useful as although the concentration of bromide at which inhibition occurs is greater than would be found *in vivo*, a concentration was found which gave good differentiation between phenotypes, especially the usual homozygote ($E_1^uE_1^u$) and the usual fluoride-resistant heterozygote ($E_1^uE_1^f$).

The order of relative inhibitions of the phenotypes is similar to that obtained with chloride, that is the atypical homozygote ($E_1^aE_1^a$) was most inhibited and the usual homozygote ($E_1^uE_1^u$) least. This was perhaps to be expected as both chloride and bromide are members of the same group in the periodic table.

When butyrylthiocholine was used as substrate with bromide as inhibitor, it was found that although bromide did inhibit the enzyme there was no concentration which differentiated the phenotypes.

Previous studies by GARRY [1971] using butyrylthiocholine as substrate showed that inhibition in phosphate buffer is greater than in Tris. The present studies with bromide confirm this. Many theories have been put forward in explanation, the main one being that phosphate itself has an inhibitory effect on cholinesterase. However, the degree of inhibition of benzoylcholine hydrolysis by bromide in both phosphate and Tris buffer is very similar. It would therefore appear that the phenomenon of postulated phosphate inhibition is only evidenced with thiocholine hydrolysis or alternatively that a reaction occurs between bromide and benzoylcholine, the product in some way preventing the inhibitory effect of phosphate.

Summary

The effect of bromide on serum cholinesterase activity has been investigated using benzoylcholine and butyrylthiocholine as substrates. A difference was found in degree of inhibition between phosphate and Tris buffer with butryrylthiocholine but not with benzoylcholine. Physiological concentrations of bromide does not inhibit the enzyme but at higher concentrations the genetic variants were differentially inhibited. Bromide numbers, that is percentage inhibition in the presence of molar bromide, give better differentiation of the usual homozygote ($E_1^uE_1^u$) and the usual fluoride-resistant heterozygote ($E_1^uE_1^f$) than is obtained with the more commonly used inhibitors. A combination of bromide inhibition with either dibucaine or succinyldicholine gave the best segregation of phenotypes.

References

ELLMAN, G.L.: Tissue sulphydryl groups. Archs biochem. Biophys. *82:* 70–77 (1959).

ELLMAN, G.L.; COURTNEY, K.D.; ANDERES, V., and FEATHERSTONE, R.M.: A new and rapid colorimetric determination of acetylcholinesterase. Biochem. Pharmacol. *1:* 88–95 (1961).

GARY, P.J.: Serum cholinesterase variants: examination of several differential inhibitors, salts and buffers used to measure enzyme activity. Clin. Chem. *17:* 183–191 (1971).

HARRIS, H. and WHITTAKER, M.: Differential inhibition of serum cholinesterase with fluoride, recognition of two new phenotypes. Nature, Lond. *191:* 496–498 (1961).

HARRIS, H. and WHITTAKER, M.: Differential inhibition of 'usual' and 'atypical' serum cholinesterase by NaCl and NaF. Ann. hum. Genet. *27:* 55–58 (1963).

KALOW, W. and DAVIES, R.O.: The activity of various esterase inhibitors towards atypical human serum cholinesterase. Biochem. Pharmacol. *1:* 183–192 (1958).

KALOW, W. and GENEST, K.: A method for the detection of atypical forms of human serum cholinesterase determination of dibucaine numbers. Can. J. Biochem. *35:* 339–346 (1957).

KALOW, W. and LINDSAY, H.A.: A comparison of optical and manometric methods for the assay of human serum cholinesterase. Can. J. Biochem. *33:* 568–574 (1955).

KING, J.: Practical clinical enzymology (Van Nostrand, London 1965).

KING, J. and GRIFFIN, D.: Differentiation of serum cholinesterase variants by succinyldicholine inhibition. Br. J. Anaesth. *45:* 450–454 (1973).

WHITTAKER, M.: The cholinesterase variants: differentiation by means of sodium chloride. Acta genet., Basel *18:* 556–562 (1968).

A. DICKSON, BSc, Royal Infirmary, Department of Biochemistry, *Glasgow G4 0SF* (Scotland)

Enzymes in Health and Disease. Inaug. Scient. Meet. Int. Soc. Clin. Enzymol., London 1977, pp. 232–237 (Karger, Basel 1978)

Enzyme Studies in Cultivated Amniotic Cells in the Field of Prenatal Diagnosis of Metabolic Diseases[1]

M. Mathieu, M. T. Zabot, I. Maire and G. Mandon

Hôpital Debrousse, Service de Biochimie, Unité d'Enzymologie et de Culture Cellulaire (Prof. J. Cotte, M. Mathieu), Lyon

For 5 years, we have been promoting prenatal diagnosis of metabolic diseases; it is a part of the activities of our laboratory for biochemical and enzyme studies in a children's hospital.

In this paper, we would like to report the problems that we met in this field, the results we obtained by studying different kinds of enzyme abnormalities and the help we got by the maintenance of a mutant cell strains bank.

The problems of enzyme studies of amniotic cells deal with methodology, sample and diseases.

Methodology

The need for rapid diagnosis requires precise enough methods of enzyme assay in very small quantities of cells. We use micro-assays for enzyme determination with 4-methylumbelliferyl (MU) substrates (fig. 1).

This procedure is not so 'micro' as that of Niermeijer *et al.* [2] who used small numbers (200–300) of freeze-dried cultured cells; but it is sufficient enough to obtain diagnosis in 3 weeks. In our studies, the results are expressed in nanomoles of substrate hydrolyzed per milligrams of proteins; this way is not very satisfactory because of the lack of precision of the Lowry micromethod. But the alternative (by number of counted cells) is not better: the different kinds of amniotic cells lead to mistakes; as reported by Niermeijer *et al.* [3] besides epitheloid, intermediate and fibroblast-like cells, some large, multinucleated cells can be present and contribute to false results when expressing enzyme activities.

[1] This work was supported by the Caisse Nationale de l'Assurance Maladie des Travailleurs Salariés.

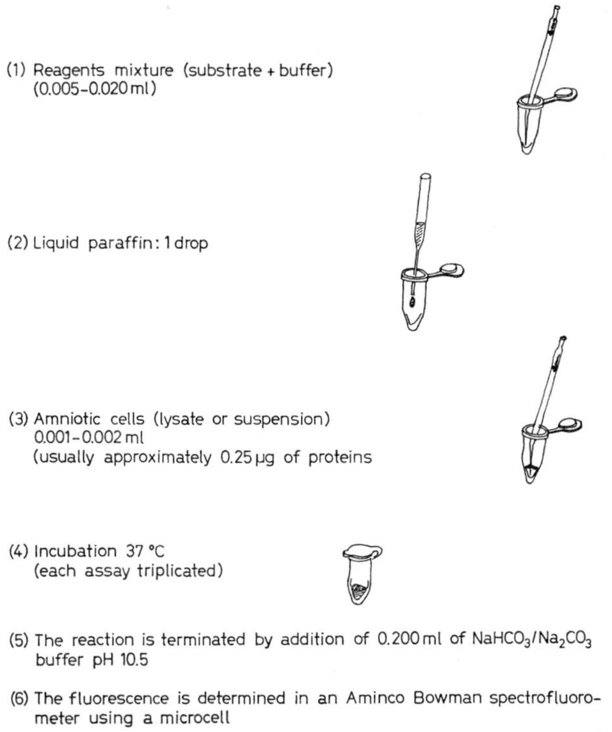

Fig. 1. Enzyme activity determination in amniotic cells with MU substrates.

Sample

The second problem comes from the sample: a large number of factors can interfere with the determination of enzyme activities in cultured cells. Because of the early stage of culture (primary or first subculture) where the enzyme activities have to be studied, it is necessary to determine reference values under the same conditions. The large variability of enzyme content during the growth of cells has to be taken into account: we are used to studying fibroblasts after the fourth subculture where enzymes are usually at their highest levels, so our reference values in these cells cannot be used for amniotic cells.

Besides, it is well known that the culture conditions have a large influence on enzyme content of the cells; for that reason it is very important to compare the tested amniotic cells with controls cultured in the same laboratory during the same time.

Furthermore, the problems common to all cultured tissues are found with amniotic cells, especially possible contaminations (bacteria, virus mycoplasma, other cells such as maternal cells...), cell density or phase of the cells in the growth cycle.

Metabolic Disease

The third problem arises from the kind of metabolic disease which has to be diagnosed: it can be an enzyme quantitative defect as in GM_1 gangliosidosis (Landing disease) (table I).

In this case, the further study of the index case had shown a complete lack of lysosomal β-galactosidase in propositus fibroblasts. Then the heterozygocity of the parents has been demonstrated; finally, the amniotic cells proved to be normal and the next newborn child is clinically normal, and this status biologically confirmed in cultivated fibroblasts (10.5 mU/mg protein – control: 10). This situation is relatively simple and frequent.

The case described in table II is simple too but more unusual: β-glucuronidase deficiency. We were able to demonstrate the deficiency in one family with two affected and two heterozygote children; then after prenatal diagnosis, a normal (or heterozygote) fetus was found on the basis of

Table I. Prenatal diagnosis for GM^1 gangliosidosis (β-galactosidase deficiency) in a single family

	Leukocytes	Fibroblasts	Amniotic cells	Amniotic fluid
Father	1.55	(1) 5.4		
		(2) 5.7		
Mother	1.04	(1) 3.4	(1) 3.4	5.3
		(2) 5	(2) 10	
Newborn	4.4			
Controls (n)	2–5 (283)	10–20 (164)	3.9–22 (20)	1.1–3 (19)

Enzyme activity: nanomoles of MU substrate/h/ml (amniotic fluid) or /min/mg protein (fibroblasts, leukocytes, amniotic cells) at pH 3.6 at 37°C.
Prefixes (1) and (2) for fibroblasts refer to data from fourth and fifth subcultures, while for amniotic cells they refer to primary culture and first subculture, respectively.

Table II. Prenatal diagnosis for β-glucuronidase deficiency in a single family

	Serum	Leuko-cytes	Fibro-blasts	Amniotic cells	Amniotic fluid
Index cases:					
Malika (A)	0.05	0.10	0.14		
Touria (A)	0.05	0.15	0.14		
Abdallah (H)	0.45	3.9			
Mena (H)	0.49	3.8			
Father	0.42	3.6	2.5	(1) 0.8	
Mother	0.42	3.9	3.3	(2) 0.9	0.5
Newborn	0.49	2.7			
Controls (n)	0.8–4.4 (51)	3.7–10 (225)	2.6–9 (142)	0.8–2.5 (9)	0.5–2.1 (11)

Enzyme activity: nanomoles of MU substrate/min/ml (serum, amniotic fluid) or/mg protein (leukocytes, fibroblasts, amniotic cells) at pH 3.6 at 37 °C.
For (1) and (2) in respect of amniotic cells see table I.

Table III. Prenatal diagnosis for mannosidosis (α-mannosidase deficiency) in a single family

Substrate	Fibroblasts		Amniotic cells	
	1 mmol × l^{-1}	7.8 mmol × l^{-1}	1 mmol × l^{-1}	7.8 mmol × l^{-1}
Index cases:				
Farida (A)	0.046	0.35		
Zakia (A)	0.050	0.47		
Father	0.20	0.51		
Mother	0.17	0.49	(1) 0.06	(1) 0.32
			(2) 0.10	(2) 0.42
Fetus (A)	0.04			
Controls (n)	0.38–1.25 (105)	0.57–1.85 (114)	0.20–1.4 (18)	0.27–2.9 (18)

Enzyme activity: nanomoles of MU substrate/min/mg protein at pH 3.8 at 37 °C.
For (1) and (2) in respect of amniotic cells see table I.

β-glucuronidase activity in amniotic cells at the primary culture (1) and first subculture (2); the newborn child has not yet been confirmed biologically.

In the third case report (table III), one of mannosidosis, we were able to demonstrate a *qualitative* enzyme abnormality in two children, then in amniotic cells and finally in the fetus after abortion. The difficulty here lies in the need for a complete kinetic study of mannosidase to diagnose the disease.

Table IV. Prenatal diagnosis of metabolic diseases in Lyon

Year	Cases	Diagnosis[1]
1973	2	maple syrup urine (A), Gaucher (N)
1974	5	Pompe's (N), Sandhoff (N), Metachromatic leukodystrophy (H), Gaucher (N), MPS I (N)
1975	3	MPS I (N), Pompe's (N), Lesh-Nyhan (N)
1976	5	homocystinuria (N), mannosidosis (A), Lesch-Nyhan (N), Menkes (N), MPS I (A)
1977	3	MPS VII (H), Landing (N), MPS I (A)

[1] N=Normal; A=abnormal; H=heterozygote; MPS=mucopolysaccharidosis.

Table V. Prenatal diagnosis of metabolic and chromosomal diseases in Lyon

Year	Total	Metabolic
1973	11	2
1974	17	5
1975	26	3
1976	80	5
1977 (6 months)	45	5
	179	18

These three examples illustrate the different situations that we can meet in prenatal diagnosis of metabolic diseases. They point to the absolute need for study of index cases before performing the amniotic cell analysis.

For this purpose, we are much helped by maintaining a mutant cell strain bank [1], where we keep the cell lines for the enzyme abnormalities which have been recognized in our laboratory, and sometimes in other corresponding laboratories. It is also important for the diagnosis of heterozygotes. Furthermore, the mutant cells are given to the researchers who are interested in the field of those inherited diseases. Table IV reports our experience in prenatal diagnosis in Lyon.

Table V gives the details of metabolic prenatal disease diagnosis in our laboratory during the last 5 years.

In conclusion, the laboratory problems of prenatal diagnosis of metabolic diseases are those of micromethods of enzyme study, cultured cell variations, and kind of metabolic diseases. Furthermore, success comes from good cooperation between different specialized laboratories.

Summary

The authors relate their experience in the field of prenatal diagnosis of metabolic disease after 5 years. They report the problems that they met concerning micromethods and enzyme determination in cultivated amniotic cells and in the help they have from a mutant cell strains bank.

References

1 MATHIEU, M.; ZABOT, M.T.; MAIRE, I.; GUIBAUD, P.; MANDON, G. et COTTE, J.: Bilan de 4 années d'expérience dans le diagnostic prénatal des maladies métaboliques: aide apportée par la maintenance d'une banque cellulaire; in BOUÉ Diagnostic prénatal, pp. 183–197 (INSERM, Paris 1976).

2 NIERMEIJER, M.F.; KOSTER, J.F.; JAHODOVA, M.; FERNANDES, J.; HENKELS-DULLY, M.J., and GALJAARD, H.: Prenatal diagnosis of type II glycogenosis (Pompe's disease) using microchemical analyses; in NIERMEIJER Prenatal diagnosis of congenital diseases; pp. 81–104 (Drukkerij Pasmans, Den Haag 1975).

3 NIERMEIJER, M.F.; KLEIJER, W.J., and GALJAARD, H.: Prenatal diagnosis of metabolic diseases in 70 pregnancies; in BOUÉ Diagnostic prénatal, pp. 163–173 (INSERM, Paris 1976).

M. MATHIEU, PhD, Hôpital Debrousse, Service de Biochemie, Unité d'Enzymologie et de Culture Cellulaire, *F-69322 Lyon Cedex 1* (France)